# TOMORROW'S WORLD
# Energy

## Malcolm Brinkworth

British Broadcasting Corporation

**ACKNOWLEDGMENTS**

There are always countless numbers of people who have helped in getting a book into print. To thank them all would take up a chapter in itself. But I would like to single out a few people: Kerry Chester, and all at the Subject Specialists Unit at the BBC who have been so patient with my seemingly endless enquiries; Pat Beales, for turning my illegible scrawl into a manuscript; to Bruce Cross, David Dancy, Richard Lord, Mike West for checking chapters; and to Annie without whom . . .

*Tomorrow's World* was first broadcast in 1965 and has been screened every year since then.

Published by the British Broadcasting Corporation 35 Marylebone High Street, London W1M 4AA

This book was designed and produced by The Oregon Press Limited, Faraday House, 8 Charing Cross Road, London WC2H 0HG

ISBN 0 563 20345 5 hardback
ISBN 0 563 20347 1 paperback

First published 1985
© The British Broadcasting Corporation 1985

Design: Martin Bristow
Picture research: Marion Pullen
Reader: Raymond Kaye

Filmset by SX Composing Ltd, Rayleigh, England
Printed and bound by Printer Industria Grafica SA, Barcelona   D.L.B. 23767-1985

FRONTISPIECE The Solar One power tower at Bastow, California, see page 61

TITLE PAGE The environmental cost of using fossil fuels

# Contents

7118503

# Introduction

When we talk of tomorrow's world many of us might think of robots, computers, advances in technology and medicine, or men pushing back the frontiers of space. We take for granted that the essential energy needed for them to operate will be there. It always has been there, at the flick of a switch, and always will be. But will it?

The last decade has seen an explosion in public awareness of the world's future energy needs. The oil crisis of the 1970s and the continuing Middle East disturbances have been strong reminders that fuel resources are finite. Our conventional fuel supplies are dwindling and one day we shall run out unless we find alternatives.

It is impossible to exaggerate the importance of energy to our modern society. It fuels our transport and heats our homes; it is the very basis of the industry on which all our economic prosperity depends.

However, for many people the energy crisis is over. It was a feature of the 1970s. No longer do energy ministers from industrialized countries feel it necessary to issue stern warnings that the world may run short of energy. As one British minister put it: 'There is no shortage of energy reserves in the United Kingdom, nor likely to be for many years'. Britain, he said, was rich in supplies of oil and gas, and had abundant reserves of coal. In 1983, for example, the United Kingdom produced almost 115 million metric tons of oil – equivalent to over 1½ times the national consumption – and 40 billion m³ of gas; 223 wells were drilled. With 'the proven technology of nuclear power available to play a more significant role', we, the public, have nothing to worry about.

As a result research into several new energy technologies was downgraded in the United Kingdom in 1983 and 1984, and it was a pattern that occurred in other countries, too. Renewable energy development now has to survive or fall according to market forces. However, the present oil glut in the Middle East, and the North Sea bonanza for countries like Britain, will not last for ever.

The problem is made worse when you think that the world population is estimated to reach between 8 and 10 billion (i.e. 8 and 10 thousand million) people by the year 2020. Half of that number, equal to the present world population, will live in developing countries on at least ten times less energy than those of us in industrial countries. Already some two billion people in the developing world are almost wholly dependent on traditional fuels, mainly firewood, to meet their domestic energy needs. It is estimated that firewood supplies are now – or may soon be – inadequate to meet these requirements in 65 developing countries. The human cost is high, too; gathering a day's firewood has grown to a full day's work. The 'firewood crisis' has become, according to the 1983 World Energy Conference Report, the 'firewood catastrophe', as forest cover diminishes globally at the rate of 250,000 km² (96,525 sq miles) per year.

So where will all this energy come from? Each chapter in the book tries to assess the size of the possible resources open to us and the state of current technology. From oil, gas and coal to nuclear power and the renewables, I have tried to look at most of the promising new developments, some of the difficulties that are still to be resolved, and some of the long-term environmental implications of what is on offer. For there is a wealth of alternative energy options and possible strategies to choose from.

One of the most modern oil platforms from the Hutton Field in the North Sea

# CHAPTER ONE
# Fossil Fuels

## Oil

Between June 1859 and the end of 1861, the quiet town of Titusville, Pennsylvania, and its surrounding countryside underwent a revolution. The town was swamped by people from all over the United States and the landscape was transformed: it was deforested, polluted, infected with the stench of sulphur and shrouded in fumes. According to one local reporter it had become the 'anteroom to hell'.

A rather dramatic effect for the success of one small project to have. But Willard Drake, a former railroad collector, had struck lucky. He had drilled the first oil well and within three months was producing ten barrels a day. In the stampede for the new El Dorado that followed, thousands of wells were drilled in Pennsylvania, and while gas and gasoline were rejected as useless, by 1862 the area was producing three million barrels a day. The oil boom had begun.

Thousands of years before in the ancient civilizations of Egypt and China, petroleum was being used for a number of purposes, including lighting. But its use as a fuel was then neglected until the early nineteenth century. Even in the European Middle Ages and Renaissance period, it was being described as a medicine rather than as a form of energy. According to one sixteenth-century writer, the oil that oozed and trickled out of the rocks near Modena in Italy was a panacea for all ills.

> First, this oil purges and cleanses any ulceration and heals all old wounds ... whosoever has been struck, disfigured or thrown so that the bruises and other effects are visible, let him rub them with petroleum twice a day and he will be cured ... he or she who has been bitten by a mad dog or has been stung by a poisonous animal, let him or her rub the wound or bite with petroleum.

Coughs, bronchitis, cramp, gout, rheumatics and eye strain – petroleum was used for them all.

Today oil is perhaps the world's most important fuel. The diversity of its products and their high value has become a major factor in the world economy. It is the source of an enormous range of raw materials that have changed the whole concept of chemical manufacture in the last century; and it is not difficult to think of a long list of essential items and processes that rely on oil for their inception and continued existence. The technological feats achieved to harness its energy have been stunning. It is claimed, for example, that the North Sea oil platforms are the largest structures ever moved across the face of the earth by man. But an oil-based economy will not last for ever; the world's supplies are finite. Laid down millions of years ago, they are rapidly being used up.

Oil was formed when the earth was mainly covered by water. As myriads of microscopic marine animals lived and died, their remains were deposited on the sea bed. Over the centuries these deposits reached greater depths and the accumulation, supplemented by plants, formed a thick slimy residue on the sea bed. Mud, brought down to the sea by rivers, mingled with the sludge and eventually turned it into oil and gas. Slowly the oil and gas were absorbed into limestone and sandstone which, because they are porous, retained the oil rather like a sponge holds water. Some deposits

Drake's oil well at Titusville, Pennsylvania, 1866

remained under the sea bed, but others became covered by land in the course of vast geological changes. Where the strata contained impervious rocks these held the oil in, creating deep reservoirs.

Once the reservoirs were found, the world set about using the fuel to its full potential. In 1984 the world was consuming oil at a prolific rate – around 27.4 million barrels a day – and the figure was even higher in the mid-1970s. So how much oil is left? Most research tends to suggest that the proven reserves amount to about 700 billion barrels, which would mean that if oil continued to be used at the same rate as it is now, it could last up to 70 years. However, our oil reserves may last considerably longer than that if the work on 'enhanced oil recovery' techniques proves successful. These techniques could bring old, 'exhausted' wells back to life, pushing the reserves up to over 1200 billion barrels; because, surprisingly enough, only between 30 and 50 per cent of the oil in a reservoir is usually recovered: the rest stays in the ground.

### A new life for old wells?

When an oil well is drilled, the impervious rock holding the oil in is pierced and the natural pressure in the reservoir allows the oil to gush out by itself. Unfortunately that pressure gradually falls and the rate at which the oil comes out declines. So for years scientists have been looking for new ways of flushing out more oil. Today water, or in some cases gas, is pumped down through separately drilled injection wells. This forces the water or gas already in the outer pores of the reservoir to push further in, expelling more oil out through the production wells. These secondary techniques work, to an extent, but to make the most of our resources all efforts are now being focused on trying to squeeze out even more.

One British oil company is turning to a compound similar to washing-up liquid for help. The additive, which alters the surface tension of the water in the way that soap and detergents do, is one of the best prospects so far for getting more out of the reservoirs. But finding the right formula is no easy task. Not only does the compound have to dissolve in the near-freezing waters of the North Sea, it also has to remain chemically stable when it reaches the deposit, where temperatures can reach 100°C. In addition, it has to keep these qualities for years and have no harmful effects on the reservoir in any way. As yet, the current experiments have not identified the magic liquid, although there are encouraging early findings.

Another entirely different approach has been to suggest that oil wells are being drilled the wrong way. For instead of drilling into the top of the reservoir, the idea is that the wells should remove the oil from the bottom. This may sound impossible, but in theory it could be done by driving tunnels under the field and then drilling upwards into the oil-bearing layers. The oil should then run out like bath water through the plughole.

However, some of the very viscous oils will need a helping hand, whatever the technique. For in some cases the stuff below ground is so thick and sticky that it is more like pitch or tar than oil. And shifting it demands rather special methods.

A new technique announced recently may hold the answer. It uses the skill that the oil industry has developed for drilling angled or curved wells, which spread out like the roots of a tree to reach the outer edges of a reservoir. In addition to a main hole, going straight down, a series of these slanted holes are drilled and each one is used to inject hot gases and steam

into the deposit. Hot solvents to dissolve the sticky oil are then pumped down the main hole.

The theory is that all these hot injections process the recalcitrant thick oil or tar so that what comes out of the production wells is a crude oil good enough to pump through the pipeline in the ordinary way. But that is not all. It is also claimed that the new process can bring out up to 80 per cent of the oil in a reservoir. If the claim proves to be true, the heavy oilfields of India, Pakistan, Indonesia and Qatar, not to mention other countries, may have a great deal to gain.

So, too, might many other areas if a project to use microbiology to squeeze the oilfields for all they are worth proves successful. The idea that a microbiologist will land on an oil platform carrying a box which could transform the world's fuel prospects may seem far-fetched, particularly when most of the industry goes to great lengths to ensure that microbial growth is kept to an absolute minimum. For most oil men fear that microbes might block up the oil-bearing strata or even the injection well itself. However, because of the work being done by Professor Vivian Moses and his colleagues at Queen Mary College in London there may come a time when the oil industry injects micro-organisms into its reservoirs as a matter of routine, and the microbiologist will be welcomed not with suspicion, but with open arms.

ABOVE LEFT The Tension Leg Platform is one of the newest designs aimed at exploiting oil reservoirs lying in deeper waters in the North Sea. It was successfully commissioned in 1984

ABOVE RIGHT A portable floating production platform which acts as a tanker

These researchers are working on a technique that could put new life into old wells using microbes. The microbes are pumped down into the reservoir. The idea is that once in place they would multiply and be ready either to manage the flow of oil through the rock by sealing off certain porous areas (thereby keeping up the pressure needed to push the oil to the production wells), or actually to convert the oil underground into a range of desirable chemical products. These might include gases like methane and hydrogen gas, as well as acids and solvents.

Work has already been carried out on a number of shallow, 'exhausted' reservoirs in eastern Europe and the United States. Innoculated with bacteria in a suitable growth medium, the wells were sealed off while the fermentation proceeded. When they were uncapped again, the wells produced an increased flow of oil. However, working in deep fields such as in the North Sea would prove a much more complex and difficult problem. Firstly, pressures in the reservoirs are very high. In the Forties Field, pressure can exceed 200 atmospheres and at Magnus it can reach 400 atmospheres. Temperatures, too, can be high, ranging from 90° to 120°C. With pressures and temperatures like that, coupled with anaerobic, oxygen-less conditions and high salinity from the sea water, the microorganisms have to be rather special to flourish. The most interesting recruits for the job at the moment come from bacteria, but it will be some time yet before the right one is found. Some of the problems may be best solved by using a group of microbes rather than by designing through genetic engineering a single super-microbe to do the whole job. But armed with the knowledge that the value of the oil which will remain in the North Sea fields will be in the order of £300 billion, you can be sure that these researchers – and others – will keep on trying.

While the oil industry is striving to make enhanced oil recovery techniques work, it is also looking further and further afield for new sources of oil and for new techniques both to find and exploit it. Working in deeper waters has meant the development of new platforms and smaller, formerly marginal fields have now become interesting commercial projects. But exploiting these smaller fields is often difficult. Most large production platforms in the North Sea, for example, are connected to many wells, each bored to tap a different part of a large oilfield. The minor pockets of oil are normally too small or too scattered to justify the expense of new platforms and they remain unused.

What has been devised is a ship that acts as a production platform and tanker at the same time. This ship, which is able to produce up to 15,000 barrels of oil a day, will receive the oil by lowering the pipe on to a wellhead already installed on the sea bottom and it will remain there until it is full. But how would such a vessel cope with the rough conditions in somewhere like the North Sea, where the wind and the waves could drive it off station, breaking the link between the ship and the pipeline in the process? It manages to remain stable because it is equipped with a series of side thrusters as well as forward and reverse propellers. These thrusters are continuously adjusted so that the forces they exert balance the efforts of the winds and currents to move the ship off station. The beauty of the system is that it is mobile, so when the field is depleted the ship can sail off to another location.

Making smaller wells commercially viable using the production tanker is one way of making better use of known deposits. But what about those deposits we are unsure about? Establishing wells that produce oil is a long-drawn-out and expensive business. A complete well costs £6 million to drill, and only about one in ten such wells looks promising enough to bring into production. And only one in ten of those actually turn out to be economic. Clearly, anything that can be done to speed up the process of deciding whether a well is going to be any good will save a lot of money.

So what about inserting 'Sniffing Pigs' down boreholes to look for oil? For that is the rather extraordinary name that chemists at the Manchester University Institute of Science and Technology (UMIST) have given a new device which will provide some idea of how hopeful a well looks. They have developed sensitive chemical sensors which measure the relative proportions of large and small paraffin molecules in the oily gases leaking into the well from the surrounding rock strata. The higher the proportion of large molecules, the better the chances of finding oil. This information is then sent back to the surface by a cable for the chemists and geologists to interpret.

### Striking oil on the city rubbish tip

While one team of UMIST chemists is working at ways of saving money on oil exploration, another group is exploring for crude oil in a rather more surprising place. Down in the chemistry department's basement an old reaction vessel is being fed with grass cuttings, used tea bags and a bizarre range of other materials to reproduce a process that took nature several million years to achieve. For the UMIST team, under Dr Noel McAuliffe, has succeeded in creating high-quality crude oil from rubbish – in ten minutes.

The process is based on the cellulose that is present in large quantities in vegetable matter, paper, cardboard, plastics and textiles from the city's dustbin. Teaming up with the rubbish tips of the Greater Manchester local authority, the chemists produced about 2.3 barrels of North Sea quality crude oil from one metric ton of feedstock. And since around three-quarters of the roughly 24 million metric tons of domestic waste produced in Britain is economically recoverable, the technique could provide a substantial amount of oil on this renewable basis.

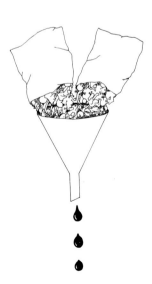

The idea of using rubbish to produce oil may seem odd. But it could prove to be very important indeed

However, perhaps the greatest impact will be felt in the developing world where huge supplies of agricultural waste matter could make up for the deficiencies and inequalities in the distribution of fossil fuels. Barbados, for example, burns the waste material left in the fields after sugarcane is harvested, yet it imports oil. If the process could be proved to be economically viable, the country could become self-sufficient in oil just using this vegetable waste.

But how is Manchester's rubbish converted into oil? Firstly the metal, glass and other inorganic constituents of the refuse are separated out. The organic material is chopped up finely and injected into the reactor, together with a simple solvent and a nickel catalyst. (Both the solvent and the catalyst are recovered and recycled.) The mixture is then subjected to temperatures of 320°C and a pressure of 7 atmospheres, and after about 10 minutes the process is complete. Nothing else is needed.

The trick lies in producing it in large quantities and in a quick and continuous process. For while other established technologies existed for doing the same thing, all of them had drawbacks according to the UMIST team. Pyrolysis, the most widely used method, tends to produce oil with a low calorific value and the reaction vessels are plagued by corrosion problems. Fermentation is another technique, but this is slow and also gives off fuel with a very low energy value. Clearly, the new method is a breakthrough and all efforts are now focused on trying to build a commercial prototype plant. So in the next few years we should find out whether all our rubbish is worth a fortune!

UMIST's way of speeding up nature to produce oil is exciting, but the process is unlikely to have a dramatic impact on the world's ravenous need for oil. Instead, many companies and scientists have been turning their attention to the vast reserves of heavy oil, tar sands and oil shale that are spread around the world. In the late 1970s and early 1980s these resources were seen as part of the bridge between fossil fuels and other forms of renewable energy, capable of lasting up to 175 years. The world's reserves of tar sands, for example, are estimated to be about one and a half times those of ordinary oil; the Orinoco region of Venezuela contains anything between 150 and 450 billion metric tons of the sticky, asphaltic oil in seams up to 30 m (98 ft) thick. Oil shale is also plentiful. The area of northwest Colorado, southwest Wyoming and northeast Utah in the United States is estimated to contain the equivalent of 8 trillion barrels of crude shale oil, although it is thought that only 400 billion barrels are likely to be economically recoverable.

However, in developing the vast deposits of both oil shale and tar sand, there would be a tremendous environmental cost. Tar sand has to be strip mined, and something like 2 metric tons of tar sand is needed to produce 1 barrel of oil. Huge amounts of water are required for the sand to be processed and there is significant air pollution. Oil shale suffers in the same way: for every barrel of oil extracted, there will be at least 1 metric ton of waste. And when plants are being planned to produce 400,000 barrels a day, that is a large amount of waste.

At the moment, most of the world's large projects planned to exploit these reserves are 'mothballed'. With the drop in the price of a barrel of oil from around $40 to its present levels, the bottom has simply fallen out of this market. It is not economic in the present climate to develop the reserves of oil shale and tar sand. And since it would take a number of years actually to exploit them when the right circumstances do arrive, it is unlikely that they will be developed in the foreseeable future.

Part of the reason for the difficulties involved in exploiting alternative or marginal oil resources is that the price of oil and its supply are now thoroughly enmeshed in world politics. The oil crisis of the 1970s forced countries to think about investing heavily in finding alternative sources of energy in order to reduce their dependence on oil. However, there is no urgency now. There is a world glut of oil and costs have fallen. Although these events may have more to do with strategic policy decisions than with the real costs of oil, all the time that the price remains stable and reasonable, countries will continue to buy oil. In present circumstances it is unlikely that they would be able to invest the huge sums of money involved in alternative energy projects.

The process of substituting other fuels for oil may take years, but the inevitable change has begun. According to the 1983 World Energy Conference, oil will drop from supplying its present 40 per cent of the world's primary demand for energy to about 20 per cent by the year 2020. So what other fuels will take over oil's role?

## Natural gas

Natural gas has always been seen as the poor relation of oil. During the early days of exploration, it was oil and not natural gas that was the objective. Gas, when it was discovered, had little practical use and was flared off in vast quantities. Even as recently as 1973, for example, Saudi Arabia flared off an estimated 14 billion m³ (18.3 billion cu yd) of gas, the equivalent of 12 million metric tons of oil. The feeling was that its commercial application was limited; unlike oil it was expensive to transport and required a substantial supporting industry.

Today, the picture could not look more different. Natural gas supplies nearly 20 per cent of the world's energy demand. It has many advantages: it is clean and easy to use and does not need elaborate manufacturing or processing; and increasingly it is seen as an attractive option both as a domestic fuel and in industrial applications.

Yet natural gas, like oil, is a fossil fuel. Its proven reserves are roughly the same as those of oil when viewed in terms of energy content. On the present estimates our gas supplies will also run out early in the twenty-first century. But the size of the world's gas resources and the conventional wisdom about where they should be could well be upset by work carried out by the Swedish Power Board, which is looking for natural gas in granite.

The search is based on the theory that much of the world's natural gas might have a non-biological origin. It could come instead from methane that was trapped deep in the earth's crust as the planet was being formed. It is a theory that scientists have been debating for years. Some, like Professor Tom Gold, believe that much of the earth's natural gas reserves are composed of such primordial gas which has migrated towards the surface of the planet. If true, that would mean that the world would not only have more gas than has been estimated, but also in different places. So far, natural gas

ABOVE LEFT Gas being flared off in a Middle East offshore platform
ABOVE RIGHT An automatic coal cutter in action

prospectors have concentrated on geological structures called sedimentary basins. Primeval gas, on the other hand, should be found in any rock fractured enough to provide a pathway towards the earth's surface and porous enough to hold a pocket of the gas.

However, nobody has been able to test the theory. And since no prospector is going to risk millions looking for primeval gas that might or might not exist, the theory is left in scientific limbo. However, the work that has been done in Sweden in the granite rock around Lake Siljan, and which could continue for the rest of the decade, represents the first steps. At the moment the studies are still chemical and geological: no drilling has taken place yet. But if the theory does prove correct, a great deal more natural gas could become available. If not, the world might turn to a substitute natural gas, made from coal.

### A new era for coal?

Early in the last century, before the advent of natural gas, coke ovens produced gas which provided domestic heating and lighting in large industrial towns in England like Birmingham and Sheffield. It was seen as the new fuel and replaced candles as the main form of illumination. As the demand grew, special gasworks were established and its use spread all over the world. This kind of gas only began to decline when electricity gained an increasing foothold, and with the discovery of the huge reserves of natural gas. But now coal is destined yet again to become a major source of gas. For the world's coal reserves are huge and are likely to last for at least another 200 to 300 years.

Most coalfields began life as swamps about 300 million years ago. In the tropical conditions prevailing at the time, plants grew rapidly and, because of the climate, decomposed slowly. Over many centuries masses of rotting vegetation built up until, in time, subsidence brought the area below sea level, burying the vegetation under mud and sand. During millions of years of being buried, sometimes beneath thousands of feet of rock, the plant remains were altered by the pressure of the rocks and by the rise in

BELOW LEFT The remnants of coalmining dominate the skyline at Blaenau Ffestiniog, North Wales
BELOW RIGHT Oil slicks cause tremendous environmental damage

temperature deep in the earth's crust. The result was coal. Different types of coal, however, were produced by different circumstances. Some plant remains experienced only mild heat and pressure, which produced brown coals and lignites; anthracite, the highest quality coal, required the extreme conditions that occurred only during the formation of mountains.

Turning coal into substitute natural gas is one of the ways in which coal will be used over the next few centuries. It is our remaining fossil fuel with large reserves, and coal technology is having a revival as a result. New designs and new processes are being developed which should go a long way to satisfying some of our energy requirements. One gasification plant being developed in Britain is a good example. The process involves feeding coal into the top of a furnace through a pressure lock. Near the bottom of the furnace, oxygen and steam are blown in under pressure, causing hydrogen and carbon monoxide gases to be produced. These are not good fuel gases but, when passed over a catalyst, they promote a reaction which results in the production of natural gas.

The advantage of the technique is that it is very efficient, with the hot gases from the reaction also passing up through the incoming coal, heating it and driving off valuable products that can go to the chemical industry. The only waste product is slag, the part of the coal that cannot be burned. It is like fine, gritty black glass and even this can be turned to good use. The scientists have found that it is good for growing plant cuttings, having the vital trace elements that healthy plant growth needs; it can be used to make a special hard coating, resistant to chemical attack and corrosion; and it could be employed to take the sulphur compound out of coal, thus cutting down pollution.

However, for one American project at Daggett, California, turning coal into a clean burning gas and chemical feedstocks was not enough. Instead they have built the first plant on a commercial scale that not only gasifies coal, but also produces power, feeding 100 megawatts of electricity into the

ABOVE AND LEFT The new 100 megawatt plant that converts coal to a clear synthetic gas at Cool Water, California

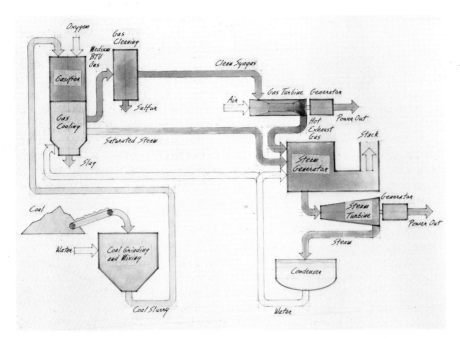

local grid. The plant, which is already efficient compared with a conventional power station, uses a 'combined cycle' to generate power costing about 10 per cent less than normal.

Combined cycle stations work by using both the gas and the steam produced from heating the coal. Gas turbines can convert heat into power at higher temperatures than are practicable in conventional stations using steam turbines. So by using gas and steam turbines together, it is possible to widen the temperature range over which heat is converted into power and make the whole process more efficient.

But to use coal in this way still means that it has to be mined out of the ground; or does it? What if the coal could be converted into gases that could be burned in power stations, or refined into higher quality fuels, or upgraded into chemicals while it is still underground? These are the virtues of a large project under way at Tono, about 130 km (80 miles) from Seattle, Washington. A team of mining engineers are drilling holes into the coal seam, igniting it and then injecting air or oxygen as well as steam into the seam to control the fire as it moves along the coal. The gases given off in the process are then piped to the surface.

The idea of burning coal in this way has been around for years, but nobody yet has properly resolved some of the main problems. One is that the coal tends to burn only along the top of the seam, leaving the lower half unaffected. This obviously results in low efficiencies, with heat lost to the rock above the coal as well as large amounts of coal left behind. But the American engineers are hopeful that they have developed a new technique that will greatly improve the efficiency of the process. Instead of drilling vertical holes into the coal seam to inject air or oxygen, as others have done in the past, they have drilled horizontally. A second hole was then added, also horizontal, right along the top of the coal seam to recover the gas and widen the burn area. They have already ignited a test area, and the only frustrating part about the experiment is that the results are not expected to be known until late 1985. Perhaps by then we shall know whether they have been successful.

Another new way of converting coal into gas has a similar 'super-microbe' feel about it to the oil wells mentioned earlier, although this microbe has very humble origins. Called DL-1, and living in sewage sludge, it could hold the key to another previously unsuccessful search – for a means of turning coal into methane biologically. Researchers at the Virginia Polytechnic Institute in the United States have found that DL-1 degrades sodium benzonate, a chemical in coal, a process which then allows its degradation products to be converted to methane by other organisms. Their studies have not solved all the difficulties yet, and there is still a lot of work left to be done. But one technology that is already available could form the real basis of coal's revival in electricity-producing plants all over the world.

### Fluidized beds

The revival may come from a process of burning coal in what is called a fluidized bed. This sounds rather jargonized, but the term really relates to putting coal into a bubbling bed of a variety of materials. Perhaps the best way to describe it is to liken it to a pan of water, only a quarter full, which is put on a cooker to boil. When the water reaches boiling point it bubbles violently and the water is rapidly converted into steam. The same principle applies to the bubbling bed, except that the water is replaced by air and the bed material consists of sand, ash, limestone and a small amount of fuel. Air is blown into the boiler and loosens up the bed material until it behaves like a bubbling liquid. When coal is added to the bed in small but constant amounts, it is burned very efficiently; couple the apparatus to a combined-cycle plant and the efficiency is improved still further.

One of the major developments in bringing about a power plant based on this technology has been under way in England at the National Coal Board's laboratories at Leatherhead, Surrey. NCB scientists have been working on a pressurized fluidized bed station which could show a significant reduction in the cost of generating electricity as well as using less coal. After years of work on early pilot plants, a full project began in January 1985 at Grimethorpe, near Barnsley in Yorkshire.

Nearly all fluidized-bed programmes run on crushed coal. But one type of fuel that is being considered not only for these plants, but also to replace oil in conventional power stations, is a mixture of coal and water. It looks a bit like thin black paint and it consists of about 70 per cent of clean, crushed coal mixed with about 29 per cent water and 1 per cent of special additives. The additives allow the slurry to be handled as if it were oil and still remain stable. It can therefore be transported and handled through pipelines, storage tanks and so on, just as easily as petroleum. But the great advantage of all of these new techniques developed for exploiting our coal reserves is that they have tackled the biggest problem concerning man's continuing use of fossil fuels. For fluidized beds and gasification processes can clean up the gases produced in conventional power plants by up to 95 per cent. And if present studies are correct, they may only just be in time.

### Acid rain

Back in 1661, John Evelyn lamented on the dirty conditions in London: 'that can never be Aer fit from them [men] to breath in, where nor fruits, nor flowers do ripen, or come to a reasonable perfection . . . It is this horrid Smoake which . . . corrupts the waters, so as the very Rain, and refreshing Dews which fall in the several Seasons, precipitate this impure vapour.' In London in 1952 thousands of people died or had to go into hospital because of the effects of this 'horrid Smoake'. A huge smog lay over the city and London came to a rather distressing halt for four days. Visibility was reduced to no more than five yards and people were confined to their homes. In 1984 over 120,000 hectares (300,000 acres) of trees in Bavaria, West Germany, were dying or dead, hundreds of Scandinavian forests and lakes had been poisoned, and common symptoms were breaking out all over Europe, and other parts of the industrial world. And the cause of all this damage lies in the burning of fossil fuels.

In Britain the great smog of 1952 resulted in the passing of the Clean Air Acts, which were meant to solve the air pollution problem. Tall chimneys were built at power stations and factories to take away the gases high into the atmosphere. However, while these measures dispensed with London's smogs, they did not get rid of the pollution; they merely transferred it. The pollution continued unabated and only now are we beginning to witness the full effects.

The pollution began in earnest two hundred years ago with the Indust-rial Revolution and it has been steadily increasing ever since. The concern is now that the earth's atmosphere has begun to reach a point where the pollutants are so large in volume that they are causing fundamental

The environmental cost of using fossil fuels is a worldwide problem

changes in the environment. The phenomenon is often referred to as 'acid rain', because the first effects that were noticed involved the acidification of lakes in Scandinavia. But the term is now generally used to refer to all sorts of pollutants, not all of them in rain, and not all of them acidic.

The principal culprits in the formation of acid rain are three gases: sulphur dioxide ($SO^2$), released by burning fossil fuels like coal and oil in power stations or factories; oxides of nitrogen ($NO_x$), which derive mostly from power stations and car exhausts; and ozone, not acidic in itself, but with an important role to play in the overall chemistry. (In Britain, 45 per cent of $NO_x$ emissions come from power stations and 28 per cent from traffic. Overall about 50 per cent of Europe's $NO_x$ emissions come from traffic.) The emissions of these man-made sulphur and nitrogen pollutants add to the concentrations of these substances that occur naturally (from marshes, swamps, volcanoes and the rotting of organic material). Worldwide, 40 per cent of sulphur compounds in the atmosphere is attributable to natural sources, and 60 per cent – equivalent to 65 million metric tons of sulphur - is emitted every year by the burning of fossil fuels and the smelting of sulphur-containing ores. However, within Europe, the balance

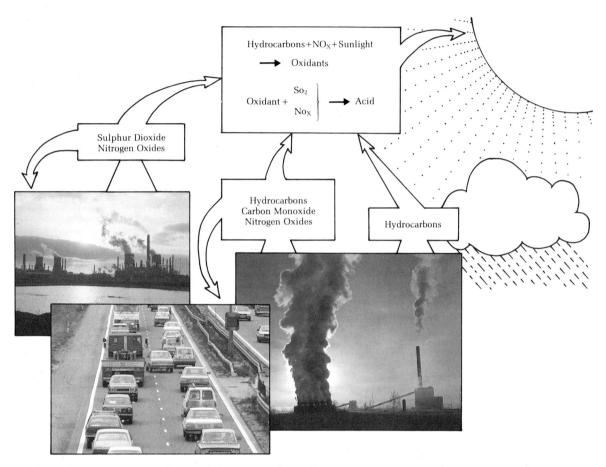

The acid rain cycle

is altered even more in favour of man-made pollutants: it produces a staggering 85 per cent. In Britain, for example, of the 4 million or so metric tons of sulphur dioxide poured into the atmosphere in 1983, 65 per cent came from power stations, nearly double the 1960 figure. In the early 1980s, the rate has actually fallen, but this is largely due to the economic recession and the lower sulphur content of North Sea oil. In eastern Europe, by contrast, more and more use is being made of highly sulphurous brown coal.

### The acid rain cycle

How sulphur dioxide and nitrogen oxides cause damage is part of a complex process. The time they can remain in the atmosphere can fluctuate tremendously, making it almost impossible to predict where and in what form the pollutants will descend. Some $SO_2$ falls directly to the ground, without reacting with the atmosphere. Known as 'dry deposition', it can corrode building materials like metal and stonework, it can affect tree and crop growths, and there is a growing fear that it may also affect human health. But when the two gases mix with the atmosphere, they come down dissolved in rain, snow, mist or fog. All these also have their effects on life.

Research at Exeter University in England has proved that sulphur dioxide and nitrogen oxide damage plants more together than their individual effects might suggest, and when combined with the chemical reaction that ozone can also cause, trees and the soil suffer as a result. Ozone naturally diffuses downwards from the stratosphere, where its concentrations may be increased two or three times by reacting with petrochemicals. And for pine trees that causes a problem. Chemicals called terpenes, given off by pine needles, react with ozone to produce highly active oxy-radicals. These then promote the reaction of $SO_2$ with water to make sulphuric acid. That can build up on the tree to release very strong acid molecules the next time it rains.

This local acid downpour is also combined with the overall acid rain coming from the chemical reactions in the atmosphere and the effect on the soil has proved to be devastating. A healthy soil is able to soak up quite a lot of acid without any adverse effects, but the defence system can only last so long. For when the acidity exceeds the neutralizing capability of the soil, its whole chemistry is affected. Poisonous metals such as aluminium, cadmium and mercury are released. Aluminium, in particular, dissolved out of soil, can damage trees. It blocks their intake of certain nutrients, stimulates the growth of many thin, short-lived roots at the expense of tough, penetrating ones, and makes the trees susceptible to disease, fungus or drought.

The cycle does not end there. The acids and metals then find their way into the nearest lake or stream, which has already been receiving extra acidity directly from rain. Unless it has some way of countering the acidity, the surface water becomes steadily more acidic. The effects are made even worse in heavy rainstorms or during the melting of vast amounts of snow in the spring, when sudden surges of very high acidity suddenly pour into the rivers and lakes. The result is that fish are stopped from reproducing and, more often than not, lakes and rivers are reduced to being a mass of clear but lifeless water.

### Only a few trees?
In June 1983, the director of the United Nation's Environment Programme summed up the situation.

> In Northern Europe, Canada, and the north-eastern United States, the rain is turning rivers, lakes and ponds acidic, killing fish and decimating other water life. It assaults buildings and water pipes with corrosion that costs millions of dollars every year. It may even threaten human health, mainly by contaminating drinking water. It is a particularly modern post-industrialization form of ruination, and is as widespread and careless of its victims, and of international boundaries, as the wind that disperses it.
> M. K. Tolba: *The State of the World Environment Report*, 1983.

The problem is getting worse. Take Germany, for example: in 1982, 7.7 per cent of its forest was reported to be suffering damage. In 1983, a more thorough national investigation was launched, with the methods of collecting data standardized in each area, and the proportion of damaged trees rose to 34 per cent. In some provinces like Baden-Württemberg, which includes the Black Forest, 49 per cent of trees were suffering. Ozone concentrations, too, have increased in Germany by more than 100 per cent in the last 15 years and other large rises have been recorded elsewhere. Thousands of lakes and rivers are now devoid of fish.

Britain, too, is not exempt from the effects of acid rain, despite the fact

A lake being limed in Scandinavia: primitive first aid

that three-quarters of its sulphur emissions are blown abroad, mainly to Scandinavia. A study recently carried out for the Department of the Environment by scientists from the Warren Spring Laboratory, to the north of London, revealed that lakes and forests are being severely polluted. They analysed data collected over the last three to four years from 38 different collecting stations throughout rural Britain and they found that acid rain was more widespread than many people had realized. At its worst, rain more acid than vinegar had fallen on central Scotland. In October 1984, the Forestry Commission also said that it was aware of 'new and quite widespread damage' affecting six species of trees, of a type not seen before in Britain.

**Preventive medicine or first aid?**
Although there is still a great deal that is not understood about the origin, transport and deposition of acid rain, some measures will have to be taken to reduce the pollution. But because of the uncertainties many scientists are reluctant to commit themselves about what can be done. They argue that it would be tantamount to a jury convicting a criminal (in this case the polluters) on circumstantial evidence alone. However, the circumstantial evidence is powerful and unless action is taken soon, the environment will be so badly affected that some effects will be irreversible. So what is being done?

As an emergency measure many of the lakes in Scandinavia have been sprayed with lime to offset the effects of the acidity. It has been partly successful as a means of giving first aid, but does absolutely nothing to solve the long-term problem. The acid emissions have to be stopped at source. Car exhausts can be cleaned up using catalysts, but only if lead is removed from the petrol first. Power stations, too, can reduce their emissions by fitting desulphurization plants which treat the gases after combustion. Japan, for example, already has 175 such plants in operation. While proposed EEC regulations would reduce $SO_2$ pollution by about 60 per cent and $NO_x$ emissions by 40 per cent, the West Germans have already acted. They have compelled 200 large power stations either to install filters to get rid of the gas or close.

Another technique, which helps remove the sulphur dioxide and nitrogen oxides, also takes place during the combustion process and is being developed in the United States. It uses electron beams and ammonia to convert 90 per cent of the polluting gases into environmentally safe agricultural products. The key to the process is that when the hot combustion gases leave the boiler, they are cooled and humidified with water and then bombarded with a beam of electrons. This ionizes the atoms in the flue gas and the ions interact with the sulphur and nitrogen molecules to form acids. Ammonia is then added to the gas stream to trigger a chemical reaction with the acids to create ammonia salt particles which can be removed and used in fertilizers.

Coal can also be 'prepared' before it ever reaches a power station boiler. This means exploiting the difference between the two main forms of sulphur that are present in coal. One is called organic; the other pyritic. The organic sulphur exists as part of the molecular structure of the coal, and because it is chemically bonded to the rest of the coal structure it is impossible to remove by purely physical means. However, the pyritic sulphur, which is predominantly iron pyrites and marcanite, can be removed. It is possible to reduce the sulphur content by up to 20 per cent, with some variation due to the characteristics of the coal and which removal process is selected.

### For a few pennies more...
All these controls on existing coal and oil plants will cost money. It has been estimated, for example, that adding desulphurization plants to Britain's major fossil fuel stations would add between 5 and 10 per cent to the price of electricity. The next generation of coal stations based on fluidized-bed combustion and gasification will not have the pollution problems. They remove up to 95 per cent of the harmful gases as part of their operations anyway. Clearly the technology exists to clean up the power stations and factories that will use fossil fuels. Whether we take advantage of them is still uncertain. But if we are to continue to use fossil fuels like coal and oil we are faced with a choice. Either we continue to pollute the environment at such a rate that irreversible damage will be caused, or we can take advantage of existing technology and get cleaner air. It is a choice that has to be made soon.

### The greenhouse effect
Unfortunately, acid rain is not the only problem caused by the burning of fossil fuels. They also produce carbon dioxide, a colourless gas, which is accumulating in the atmosphere. According to most scientists, the ever-increasing level of carbon dioxide could change our climate dramatically. In Britain the Royal Commission on Environmental Pollution warned at the

beginning of 1984 that the potential dangers are so great that we should start planning for a possible switch away from fossil fuels towards other sources of energy. In the United States two well-respected bodies, the Environment Protection Agency and the National Academy of Sciences, suggested that the world may be warming up at an ever-increasing rate. Their vision is that the earth's temperature will rise, creating a steam bath, and along with this the flooding of low-lying areas and vast climactic changes that may make crop-growing impossible in many regions where it is of vital importance today (such as the great grain belt of the American Midwest).

These rather dramatic predictions are based on years of research and there is now little doubt that carbon dioxide levels are rising. A continuing increase also seems unavoidable at least for several decades and, if present trends continue, within a hundred years the overall level is likely to double. This would mean that the temperature could rise by 2° or 3°C. This is because carbon dioxide lets sunlight through to the earth, but prevents the escape of low-temperature heat and radiation. It acts much like the glass windows in a greenhouse, causing the earth's climate to warm up in the same way that the inside of a greenhouse does.

What the two American reports suggest is that this warming – at first, perhaps, only a degree or two eventually reaching up to 5°C – would melt polar ice, so raising sea levels. While coastal regions might have more rain, contaminated areas would change the map of the world. The changes will be gradual, but the worst results of the greenhouse effect will hit the world by the twenty-second century.

In the light of the impending carbon dioxide doom, perhaps the only answer is to use less fossil fuels, or not to use them at all. Since for the immediate future we cannot do without them, could we be more efficient with what we already use? Could we make use of waste heat or power to at least reduce the likelihood of a global steam bath?

**Conservation**
Most of us, thanks to the glories of modern central heating, can live in warmth and comfort even on the coldest winter nights. Whereas a couple of generations ago people commonly accepted the need to dress according to the season, today we can walk around in a T-shirt even when it is freezing outside, simply by turning the heating up. But although our buildings have become massive fuel users, modern techniques have also made them more energy efficient. Houses built to the highest thermal standards that are economically possible today need only half as much fuel for heating as their equivalents built to current British building regulations, and a tenth as much as the pre-1930s equivalent with solid walls. However, the savings are not just part of a 'Save It' campaign to get everyone to use less. They form part of an overall strategy for all our industries, goods, buildings and technological processes to make the most efficient use of energy.

The potential fuel that could be 'saved' in Britain alone is a good example of what could be achieved on a much larger scale. It has been estimated that up to 30 per cent could be saved in industrial processes and industrial buildings; over 40 per cent in domestic, commercial and public buildings; and 20 per cent in transport. This sounds rather over-optimistic in the short term, but there is widespread agreement that savings of around 20 per cent or so are economically and technically feasible right now. That would mean that of the £100 million a day that Britain spends on energy, £20 million a day of this is wasted. The value of a 20 per cent saving would be worth around £4 billion a year (at today's prices) by the end of the century.

### Combined heat and power

Another form of energy that is simply going to waste is the heat produced at power stations arising out of the generation of electricity. At the moment huge towers dissipate this 'waste', so that it heats the clouds rather than being used in buildings or in any industrial processes. There are a number of combined heat and power (CHP) stations working in Britain, but they are generally few and far between. For a technology that was first installed at the sugar factory in Clydebank, Scotland, in 1898, it has had a remarkably slow development.

One of the great difficulties was the creation of the national grid. This allowed the concentration of fewer, larger power stations, sited away from centres of population. Any use of all the spare heat had then to be near the station, rather than where the demand was. But now, CHP plants coupled up with district heating systems look both economically and environmentally attractive. So, too, do schemes for industries that have a steady demand for both low-grade heat and electricity. The main reason is that previously successful CHP schemes have always required a very fine balance between the demand for electricity and the need for heat. Quite often the two were incompatible and the station would then be uneconomic. With the passage of the 1983 Energy Act, however, the old equation does not apply in Britain any more, because any fluctuations in the demand for electricity can now be accommodated by the normal electricity network. So if the station is producing too much electricity because more heat is wanted, it can sell that off to the national grid; if it is producing too little it can buy it at the standard rate. The British national grid, once a disadvantage to CHP stations, has now become a kind of power reservoir; and this time everyone is hoping that CHP will flourish.

Although big stations may only just be entering their 'Renaissance' period, the development of smaller compact units has already been successful. In West Germany, for example, over a hundred small CHP plants have been installed and results from the country's energy research programme look very encouraging. They show that it is possible to save 50 per cent of the primary energy required for heating purposes, when the heat supply is accurately adapted to the local housing conditions.

A great deal of the energy-saving breakthroughs that are occurring are possible because of the introduction of modern electronics. It is inevitable that sophisticated electronic controls will play an increasingly important role in managing the use of energy in buildings and in manufacturing processes. By the early 1990s it is likely that such controls will be helping electricity companies regulate the temperatures in homes and buildings in line with how much power they need, enabling them to reduce spare capacity on the grid, so saving energy and money.

However, all the sophisticated technology that may help us to reduce our energy consumption will not allow us to get away from some inescapable facts. Our fossil fuels will run out, and they will continue to have some very undesirable effects on the environment. So what will be powering tomorrow's world? With no sulphur or nitrogen oxide emissions, with no slagheaps, or oil spills, perhaps the world has a cleaner alternative in nuclear power. So can the atom be made to work for us?

# CHAPTER TWO

# Making Atoms Work

On 2 December 1942, a group of physicists gathered in a disused squash court under Stagg Field football stadium at the University of Chicago. Shaped like a door knob, a layer of solid graphite blocks stood on a wooden frame, in some of which were embedded balls of uranium metal or compressed uranium oxide powder. The pile was 51 layers thick and the whole structure was 8 m high (20 ft). A series of cadmium rods stuck out of the top, which could be moved up or down by links on the control panel.

At the panel, Professor Enrico Fermi directed the controls until one of the dials flickered and an instrument began ticking. Smiling, Professor Fermi turned to his audience: 'Gentlemen,' he said, 'the chain reaction is self-sustaining.' And with that statement the world's first nuclear reactor was born.

The Chicago squash court was not perhaps the most impressive of settings for nuclear fission to make its momentous début. But the effects of Fermi's demonstration certainly have been impressive. For nuclear reactors, harnessing the power of the atom to produce energy, have been hailed as the heart of the technology that may shape tomorrow's world, or may obliterate it. Nuclear technology brings with it the fear of the unknown, with its power to emit harmful radiation which we can neither smell, taste nor touch, killing or mutilating for generations; it also can be tamed to provide a virtually inexhaustible source of energy. Since our fossil fuels will not last for ever, can we afford not to make atoms work for us, exploiting the power of the nuclear pile?

## Splitting the atom

Fermi's demonstration was the culmination of years of research aimed at understanding the physics of the atom, and it rested on two important

BELOW AND LEFT Enrico Fermi and the Squash Court Reactor

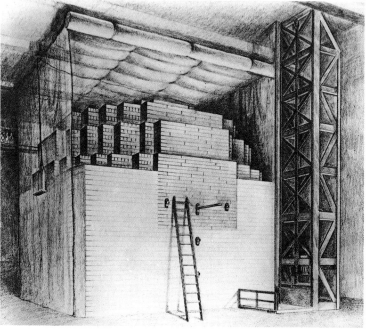

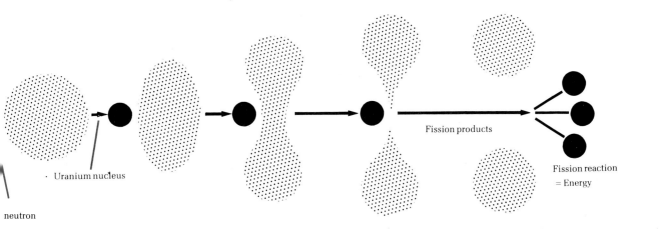

neutron

· Uranium nucleus

Fission products

Fission reaction
= Energy

The fission reaction

discoveries. The first was the existence of the neutron. By the 1920s, it was known that atoms contained positively charged protons inside a nucleus and negatively charged electrons flying about around them. But no known force could hold together lots of particles, all with a positive charge, in such a tiny nucleus. Electrons, it was suggested, might be acting as a sort of nuclear glue – but the theory did not stick. Some kind of heavy particle, which carried no charge, had to be the answer: it was the neutron.

The second step was crucial and proved to be perhaps the most fateful discovery of the twentieth century. Otto Hahn and Fritz Strassman, working in Berlin in 1939, found that they could make heavier elements by bombarding uranium with neutrons. But in the process of doing so, they also accidently discovered that a uranium atom could be divided into two parts, generating a colossal amount of energy as the atom split apart. For as the uranium atom separated a small amount of matter was destroyed; and because energy and matter are interchangeable (Einstein's famous equation $E = MC_2$), it was this small amount of matter that was converted into the most concentrated and powerful energy source ever known. It was called nuclear fission.

Uranium is the only naturally occurring element in which fission will happen easily, and an individual atom only needs one strong neutron blundering into it for the reaction to take place. The uranium nucleus, however, contains many neutrons and, when split, it not only releases energy, but also two or three more neutrons. Flying off in all directions at tremendous speed – about 16,000 km (10,000 miles) per second – the neutrons can rupture further uranium atoms, causing more fissions and releasing more neutrons. If there are enough uranium nuclei together, the atomic havoc multiplies extremely quickly. As more and more neutrons are released, more and more ruptured nuclei occur, with more and more energy being generated: and the process continues until there is a chain reaction.

**The chain reaction**

The great skill Professor Fermi showed on the Chicago squash court was to obtain that controlled, self-sustaining chain reaction, which supplied itself with the right number of neutrons. For to keep the chain going each neutron that is lost by causing a fission has to be replaced by exactly one neutron that does the same. The problem is that not every neutron splits another atom. So each time a neutron misses its target, the chain reaction is weakened.

The delicate balance of neutrons that has to be achieved is not helped by the composition of the uranium itself. Only 1 in 140 of the atoms in natural uranium are of a type that will fission easily when hit by a neutron.

Natural uranium atoms come basically in two forms, known as U235 and U238. They are chemically identical, but U238 has 3 extra neutrons and for fissioning the difference is crucial: 99.3 per cent of natural uranium is made up of U238 and only 0.7 per cent is U235, but it is the U235 atoms that fission.

During the neutron bombardment of a lump of uranium, the neutrons move at different speeds. The U235 atoms fission when they are hit by slow-moving neutrons, but the U238 ones do not. They merely gobble them up, without fissioning, and reduce the number available to their U235 neighbours. Ideally the reactor would have only very slow neutrons moving about in a pile. For physicists, therefore, the challenge was somehow to improve the prospects of a chain reaction taking place. To do that, they had either to increase the proportion of U235 to U238 in uranium – to 'enrich' it – or to slow down the fast-moving neutrons to speeds at which they could be absorbed by the U235 atoms. Alternatively, they could do both.

The best way to slow down, or 'moderate', the speed of the neutrons is to surround the uranium with a material that is able to do just that but without absorbing the fast neutrons in the process. The neutrons would simply bounce around like a ball on a snooker table until enough energy had been lost. The best materials for the purpose are graphite and water. But having got the chain reaction going, how do you control it? The answer lies in materials that do the exact opposite of the moderator; they soak up neutrons like a sponge absorbing water. These absorbers are rods placed in the reactor's core and are usually made out of cadmium or boron. When they are in place, they prevent any chain reaction taking place, absorbing all the available neutrons. However, by lifting them slowly out of the core, stage by stage, the neutrons are freed and the fission reaction begins. Perhaps the best way of describing their role is to compare them to an accelerator pedal on a car. The more you open up the throttle the quicker you go; ease off and you slow down. Lifting or lowering the rods has the same effect on the chain reaction, keeping it under control.

To be able to create a controlled, self-sustained chain reaction was the master key to the harnessing of nuclear power. While Fermi and his colleagues on the Manhattan Project went off to exploit the power of uncontrolled fission in the atomic bomb, peaceful exploitation of the atom was established. The last 40 years or so since that basic discovery has seen the expansion of nuclear power stations all over the world. There are something like 300 units now operating and still more under construction. They all tend to fall into three basic categories, determined by how they are 'cooled' or extract the heat produced from the reactor. Some are gas cooled; others use water. However, they all share the same basic fuel, uranium, and

The fuel pin being loaded at Sellafield, Cumbria, ready for the Prototype Fast Reactor at Dounreay, Scotland. The fuel comprises mixed plutonium and uranium oxide pellets in stainless steel cans

**Comparison of different types of reactors**

| Reactor type | Moderator | Coolant | Coolant outlet temperature | Coolant pressure | Uranium enrichment efficiency | Steam cycle efficiency | Core dimensions (dia×height) | |
|---|---|---|---|---|---|---|---|---|
| Magnox | Graphite | Carbon dioxide gas | 400°C | 300 psia | Natural uranium (0.7%) | 31% | 14 m × 8 m | Data based on 600 megawatt size |
| Advanced Gas-Cooled (AGR) | Graphite | Carbon dioxide gas | 650°C | 600 psia | 2.3% | 42% | 9.1 m × 8.5 m | Data based on 600 megawatt size |
| Pressurized Water (PWR) | Ordinary water | Ordinary water | 317°C | 2235 psia | 3.2% | 32% | 3.0 m × 3.7 m | Data based on 700 megawatt size |
| Boiling Water (BWR) | Ordinary water | Ordinary water | 286°C | 1050 psia | 2.6% | 32% | 3.7 m × 3.7 m | Data based on 600 megawatt size |
| CANDU | Heavy water ($D^2O$) | Heavy water ($D^2O$) | 305°C | 1285 psia | Natural uranium (0.7%) | 30% | 7.1 m × 5.9 m | Data based on 600 megawatt size |
| Steam-generating Heavy Water (SGHWR) | Heavy water ($D^2O$) | Ordinary water | 272°C | 900 psia | 2.24% | 32% | 6.5 m × 3.7 m | Data based on 600 megawatt size |
| High Temperature (HTR) | Graphite | Helium gas | 720°C | 715 psia | 10% | 39% | 9.8 m × 6 m | Data based on 1300 megawatt size |
| Leningrad (RMBK) | Graphite | Ordinary water | 284°C | 1000 psia | 1.8% | 31.3% | 11.8 m × 7 m | Data based on 100 megawatt size |
| Fast Reactor Reactor | None | Sodium | 620°C | 5 psig | 20% plutonium | 44% | 2.3 m × 1.1 m | Data based on 1300 megawatt size |

produce electricity.

### Too cheap to meter?

When nuclear power first appeared on the public scene, it was accompanied by a fanfare of predictions and publicity. They ranged from the sublime to the ridiculous. 'Atomic power', as it was called, would run a car on an engine the size of a fist; we would soon live in houses heated by uranium; 'atom-powered' rockets would enable us to cross oceans in 3 minutes. It all sounded fantastic. Electricity would almost be too cheap to meter if you listened to one report; another version, more conservatively, put the cost at one-fifth of any fossil fuel.

While no one now believes that thermal nuclear power is the answer to the world's energy problems, optimism has surrounded the development of a different kind of reactor – one that literally breeds its own fuel.

### The fast breeder

All conventional nuclear reactors work by slowing down the fast neutrons from the fission reaction until they have the same energy as the surrounding atom. They are then in thermal equilibrium and as a result, the reactors are known as 'thermal reactors'. They share one major common feature. Whether the fuel has been enriched or not, it has to be replaced every few years. It loses much of its original stock of easily split nuclei and at the same time it acquires a large stock of neutron-absorbing products. But one product that is produced actually helps prolong the chain reaction, and that is plutonium.

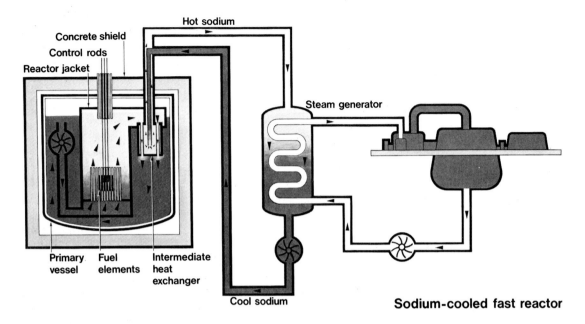

**Sodium-cooled fast reactor**

Plutonium, an element that does not occur naturally, is created when U238 atoms absorb some of the available neutrons in the reactor. And like its U235 counterpart, plutonium (239) is fissionable and can make a significant contribution to the total release of energy. What scientists, therefore, have been working on since the late 1940s is to produce a reactor that can utilize plutonium as a fuel and use the fast neutrons. For whereas the thermal reactors use up only a very small proportion of their fuel – most of it is wasted or useless, and has to be removed and reprocessed – plutonium is a very efficient fuel source. However, it needs a very different design if it is to be utilized.

The fast 'breeder' reactor has enormous potential. It enables 60 times as much energy to be extracted from a given quantity of uranium as today's thermal reactors can. It can do so because it utilizes spare fast neutrons to convert U238 to plutonium 239, which can be extracted and then used as fuel. In a thermal reactor, slow neutrons or moderators are used to help them collide with U235. But by using 20 per cent plutonium, no moderator is needed.

But how can a fast reactor 'breed' its own fuel? A fission reaction caused by a fast neutron produces on average more new fast neutrons than a thermal neutron. Some are used up, maintaining the chain reaction; yet a lot of surplus neutrons do escape from the core. But by placing a blanket of the relatively worthless U238 around the core, they can be absorbed. In doing so, some of the uranium is changed into plutonium, which can be 'harvested' to be recycled as a new fuel in the reactor core. The rate of plutonium production can vary, but after about 30 years, a fast reactor can create as much plutonium as it consumes. Fast reactors would enable huge amounts of energy to be produced from the poorest source of fuel, the U238 atom, which makes up 99 per cent of natural uranium. It sounds tremendous (a self-sustaining chain reaction producing fuel on a renewable basis). Unfortunately there are one or two problems.

Since research began nearly 40 years ago many experimental fast reactors have appeared on the energy scene and been operated. Most of them have been small, with three intermediate prototypes emerging out of the nuclear cast. The British prototype development has been centred at Dounreay in Scotland, where two stations have been constructed. The first, which operated between 1959 and 1977, could produce 15 megawatts of electricity. It was then shut down, being superseded by the second and much larger 250 megawatt prototype. This began operating in 1975 and in 1981 the first fuel from the fast reactor was reprocessed, remade into new fuel elements and returned to the reactor.

However, the development of the fast reactor into a commercial prototype is still a long way off. Although British researchers have been at the forefront of the technology – for example, in trying to make their reactor consume more than 10 per cent of the fuel before it has become so changed by conditions inside the reactor that it has to be brought out for reprocessing – it will be years before the fast breeder is economically viable, Other work is continuing in France, with the construction of a 1200 megawatt Super-Phénix station at Creys Malville, and in the Soviet Union the BN 600 design is already operational.

However, with the present annual expenditure in Britain at around £100 million, power from the fast breeder is not going to be cheap when we eventually get it. A programme of breeder reactor construction is unlikely to be needed or justified on economical grounds for at least a few decades. The Central Electricity Generating Board, in its evidence to the Sizewell Inquiry, stated that 'it might need to order fast reactors on a commercial scale in preference to thermal reactors in about 25 years from now' – but in reality it could well be earlier or later. No one really knows.

Dounreay site, Caithness

One vision of the future of the fast breeder has already been publicly aired by the United Kingdom Atomic Energy Authority. Dr Tom Marsham, the Managing Director of its northern division, foresees four parks of fast breeders in the next century – each containing eight reactors, totalling up to 10,000 megawatts – which would provide about a half of British electricity needs. They would require no fuel for hundreds of years, dealing with their own wastes, and would be a totally reliable electricity source. Sir Alan Cotterell, former Chief Scientific Adviser to the British Government, has suggested that in the twenty-first century 'there could be a few thousand power stations in the world, mostly of the fast breeder reactor type, all running on plutonium fuel. To service these, some 50 to 100 reprocessing plants could be needed and over 30,000 tons of plutonium.' If all this were to come true, our energy supplies would have become dependent upon the availability of plutonium and we would have moved into the plutonium economy.

The plutonium economy is an idea that has frightened many politicians and energy scientists for years. Dependence upon an element with tremendous potential for destruction and intense radioactivity confers a level of responsibility on its users which we are only just beginning to understand. Fast breeder reactors may offer a renewable energy source, providing power in the twenty-first century, but they also require a level of vigilance over their products of military proportions.

However, nuclear engineers and physicists have also been working on another potential source of power from the atomic stable since the early days of nuclear research, one that has been heralded many times as a renewable energy source that could ensure mankind's power supply for the next 100,000 years, with fuel costs that would be almost negligible. It is nuclear fusion.

## Fusion

The world of predictions is a dangerous one. Many well-respected scientists who have looked forward into tomorrow's world have been proved totally wrong. Inventions that were going to change the world have faded ignominiously away, while others that seemed destined for the oblivion of some learned tome have turned into great successes. Charting the progress and future of fusion power is laden with all the same risks. But many nuclear scientists are hoping that it may be the nuclear energy of the future.

Fusion has already had some false dawns. As far back as 1958 newspapers were telling us that by 1978 we would have no electricity bills to pay and a free energy bonanza would last for ever. It was all thanks to 'the miracle of nuclear fusion, the awesome power of the H-bomb, tamed for peace'. The excitement and optimism of the 1950s soon disappeared as experiments revealed that the process was more complicated than had been thought and the goal of controlled fusion seemed further away than before. A worldwide programme of research and development was inaugurated with Britain and other European countries, Russia, and the United States taking the lead. However, the day of the fusion reactor still remains many years away.

But what is fusion? The nuclear power plants that we use today work because of the fission reaction. Fission, you will remember, means that the nucleus of a heavy atom such as uranium is split to form two smaller ones; and it is done by using one of its component parts, a neutron. Fusion does the exact opposite. Instead of smashing a heavy nucleus apart, the idea is to take two small nuclei, deuterium and tritium, and force them together to form a heavier one. When the two nuclei, which are actually different forms of hydrogen isotope, are fused together enormous amounts of energy are released. It is a reaction that takes place naturally on the sun and it would effectively amount to creating a man-made sun under controlled conditions on earth.

The great advantage of fusion is that it produces far more energy than conventional nuclear fission, and there is no radioactive waste to worry about at the end. It is not totally benign, in that tritium itself is radioactive and high-energy neutrons released from the fusion reaction would make the reactor structure radioactive; but the only waste product of the reaction itself is the inert gas helium. It needs no reprocessing facilities and its fuel supply is limitless. Fusion would resolve the future of the world's energy needs: there is all the deuterium we would ever need in the sea – it is a component of ordinary water – and tritium is readily available by making it from lithium. (Tritium is in fact made by surrounding the fusion region with a blanket of lithium. This slows down the neutrons from the fusion reaction, producing heat and reacting again to breed fresh tritium. It is radioactive and has a half-life of 12.3 years.)

### Plasma
Nuclear fusion has already been achieved with the explosion of the hydrogen bomb. In the H-bomb energy was released in an uncontrolled and destructive reaction. The challenge ever since has been to reproduce on earth the events that generate the immense energy of the sun under controlled conditions to produce electricity: and it all depends on being able to create plasma.

In a gas at normal temperatures atoms are neutral: they are balanced out with a positively charged nucleus and negatively charged electrons which orbit the nucleus. But if the temperature rises, the electrons effectively reach escape velocity and fly off. So the natural gas is transformed into a collection of charged particles, called an ionized gas, which is influenced by electric and magnetic fields. This condtion is called a plasma – the sun and the stars are all in a permanent plasma state.

The sun, a giant fusion station, manages to keep its reaction going as gravitational forces hold all the changed particles together. On earth, however, trying to force two positively charged nuclei into one element is extremely difficult because, just like two similarly charged magnetic poles, they repel each other. To get round that problem the nuclei must be travelling so fast that they can overcome the electrical repulsion and fuse together. One way of achieving this is to heat the mixture of deuterium and tritium gas to a cool 100 million degrees C.

### Heating the plasma
In order to produce more energy out of a fusion reaction than the vast amount of energy that is put into it, there are two very difficult things to be mastered all at once. Firstly, a way has to be found to create that incredible temperature of 100 million degrees C. At the moment it is possible to create routinely 30 million degree plasmas for up to one second. Larger experimental devices are coming into use and it is hoped that they will be able to create plasmas with temperatures between 50 and 100 million degrees

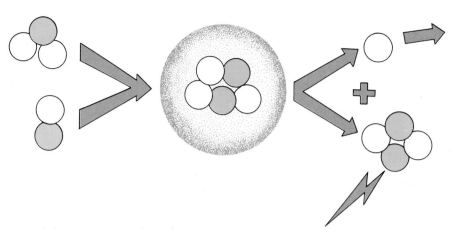

LEFT The fusion reaction: tritium and deuterium becoming helium

for tens of seconds, long enough for abundant fusion reactions to occur. The highest temperature recorded so far has been at the US experimental reactor at Princeton, New Jersey, which reached 80 million degrees C.

But how do you actually generate such heat? Since plasma is a good conductor of electricity it can be heated by passing an electric current through it. However, the current has to be huge, measuring millions of amperes. At Culham, Oxfordshire, where the Joint European Torus (JET) is housed, currents of 3 million amps were produced at the end of 1983, an achievement the project manager described as an 'early milestone' in the drive towards the full load of 4.8 million amps.

However, just heating the plasma with an electrical current will not reach the high temperatures needed, because it gradually loses its ability to be heated by the current as the temperature rises. Working at the moment with

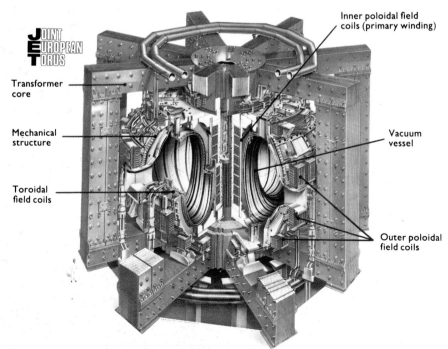

Inner poloidal field coils (primary winding)

Transformer core

Mechanical structure

Toroidal field coils

Vacuum vessel

Outer poloidal field coils

LEFT AND OPPOSITE BELOW The Joint European Torus (JET) at Culham, Oxfordshire OPPOSITE ABOVE To reach the enormous power required to make laser fusion work, the beam has to be amplified. But as it becomes more powerful, it must be spread out over a larger surface to prevent damage to the glass dishes. This is achieved by using spatial filters, which work like telescopes in reverse

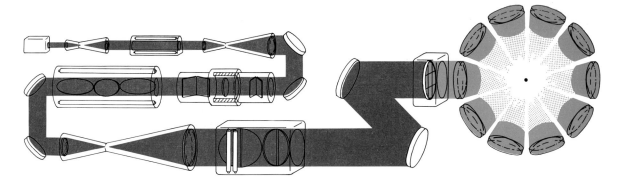

just hydrogen as the plasma material, and not deuterium and tritium, the next phase at Culham is to squirt very high speed hydrogen into the torus. For when the hydrogen atoms strike the plasma they will share their great energy with the plasma nuclei, heating it up further. By this means the scientists hope to reach higher temperatures; but they must resort to yet another device, called radio frequency heating, to raise the hydrogen up to the 50 million degrees mark. Using radio frequency waves to heat the plasma the process is rather like the one that goes on in a microwave oven. It

LEFT The interior of the vacuum vessel at JET. It was made as a totally welded structure to ensure that the required high vacuum conditions can be achieved. The chamber has double skinned walls

still falls short of the magic 100 million, but if all three stages can be made to work well, deuterium and tritium isotopes will be used for the first time. Then, it is hoped, the scientists will be able to reach what seemed to be an unattainable target.

### Confining the plasma

Having heated it up, the second crucial stage is finding a means of keeping the plasma from flying apart. The most favoured method of confining the

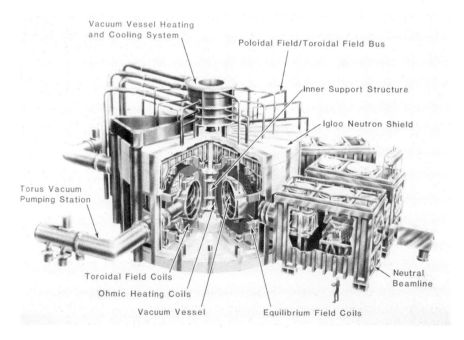

LEFT The US experimental reactor at Princeton, New Jersey

plasma, and the one employed at JET, is to use magnetic fields. Just as gravity causes a satellite to orbit the earth instead of shooting off into space, so a magnetic field forces a charged particle trying to cross it to follow a spiral path. By surrounding the plasma with magnetic coils, the field produced forces it away from the container walls. This is crucial because contact with the container cools the plasma down and stops the reaction.

To ensure that the plasma does not leak out of the ends, the reactor is ring-shaped, or toroidal. It acts as a sort of vacuum flask. The design is called the Tokamak and it was developed in the Soviet Union in the 1960s. Early experimental results soon established its pre-eminence and the Tokamak design is now the front runner for fusion systems.

### Laser fusion

However, although the Tokamak may well be the front-running design, the use of magnetic fields is not the only way of confining the plasma. Alternative designs flourished before the emergence of the Soviet scheme and they have continued to do so. In both the United States and the USSR scientists are working on a method that uses lasers. The idea is to take a small hollow pellet, about 6 mm (¼ in) across, and put the deuterium-tritium fuel into it when the mixture is cold and easily controllable. To get it hot enough for the fusion to take place, the pellet would then be bombarded by powerful laser particle beams. The beams will not only heat the fuel up to the right temperature, but will also boil off the outside of the pellet forcing it to collapse on itself. This fuses the nuclei before they have time to fly apart.

There are problems with the laser technique, but the nuclear engineers involved believe they are not insurmountable. One is that the energy released from each pellet is tiny and, unlike fission, the fusion reaction in the pellet does not trigger off a self-sustaining chain reaction. To get more energy out of the reactor, the whole process has to be repeated and repeated – about a million times a day.

A second problem is that lasers are themselves great users of energy. The most efficient laser today converts only five per cent of the power needed to produce its beam into actual beam energy. So to achieve the target of a commercial fusion power station getting more than you put in using lasers is a difficult task.

An alternative to the laser is to use beams of positively changed ions to compress the heat fuel pellets. This is a technique that is being followed up in Britain and in West Germany in a collaborative project, and the first experiments were carried out in late 1984. The advantage of the ion beam is that it converts into energy 30 per cent of the power that is needed to create it. It makes it a much more efficient and economic option than lasers.

However, a big question mark still remains. Nobody knows yet whether it will be possible to focus intensely energetic beams of particles which naturally repel each other on to a tiny target and then hold the beams focused there. The magnets used for focusing would have to be so powerful that they make the whole idea too expensive: they would have to handle an amount of energy equivalent to all the sunshine falling on the whole of Britain on a hot summer day, focused on a tiny pellet one millimetre in diameter.

Accomplishing all the challenges that face fusion power at the moment may seem a distant dream: huge magnetic fields, temperatures hotter than the sun, reactors that can cope with a million small explosions every day, and lasers that will have to be ten times more powerful than any yet built. The goal of cheap, clean nuclear power is not just around the corner by any means. The scientists still have to prove that a thermonuclear reaction could generate enough energy at a low enough cost to become the furnace of a power station boiler. And cost may well be a crucial factor.

At the moment the world is spending about $1 billion a year on fusion research and development. This investment is to explore the basic physics of fusion reactions, using applications like JET, and not to develop the technology and engineering of power reactors. The aim is that by 1989 the plasma will have been heated to the 100 million degree C mark and fusion supporters hope that by 2020 the first large-scale reactor will be ready for use. The fission programme is not without its critics. The concern is that the cost of harnessing such powerful reactions safely and of frequently replacing expensive reactor parts destroyed by the intense radiation would put fusion power beyond the cost of fast reactors.

The scientists believe that fusion will be worth the wait if they can overcome their difficulties. For 1 m³ (35.3 cu ft) of water, which normally contains 1 deuterium atom for every 6500 hydrogen atoms, has a potential fusion energy of 8,160,000 million joules or the heat of combustion of 269 metric tons of coal or 1360 barrels of oil. Fusion could prove to be the solution to the world's energy problems – or at its worst prove a very expensive alternative. By around 2020 we should know.

BELOW AND OPPOSITE Nova – the world's most powerful laser – which is being used to develop research into nuclear fusion. These two pictures show the chamber where the fusion reactions will occur and the source of all the power generated: the laser room

# CHAPTER THREE
# A Radioactive Future?

Over the last few years a new kind of traffic has appeared on the roads, railways and sea lanes of Europe. Its destination is one of two large coastal factories, at Cumbria in England and Normandy in France, and the cargo is one of the most dangerous substances known to man. Encased in up to 80 metric tons of steel plating – tough enough to withstand a high-speed train crash – it is spent nuclear fuel being transported from over 100 nuclear reactors in western Europe and Japan. It is a traffic that has aroused strong public and environmentalist reactions, and is more than just an indirect consequence of nuclear power; it is an intrinsic part of the second phase of the nuclear fuel cycle. For every nuclear reactor produces radioactive waste and fuel that needs to be reprocessed.

Compared to fossil fuels, nuclear power seems to offer a clean, compact alternative. It produces no waste gases to increase the level of carbon dioxide and sulphur in the atmosphere, nor does it produce millions of tons of ash. Instead, the end of the nuclear power cycle creates small quantities of concentrated radioactive substances, some of which will remain potentially harmful for thousands of years. It is these wastes, and people's fears about the whole subject of radiation, that have prompted so many questions about the safety of nuclear reactors, and how the radioactive substances we are left with will be safely disposed of and stored.

In many European countries this concern has been turned into a political issue. In Austria, for example, a referendum decision halted any further nuclear plans. In Sweden the nuclear question became so political that it caused the downfall of two governments, and in March 1981 a national referendum was held. A policy calling for the phasing out of all nuclear power in Sweden by 2010 won a majority, despite the fact that Sweden has 70 per cent of Europe's uranium deposits, and nuclear power produces around 40 per cent of the country's electricity.

In contrast other countries and governments have continued to expand their nuclear power programmes, believing that nuclear power is a vitally important technology and that the problem of managing radioactive waste is not insuperable. But what is nuclear waste and how is it formed?

After a period of time in the reactor, the uranium fuel becomes less reactive and has to be replaced. However, the old fuel still remains a very valuable source of energy. Only about half of the fissionable uranium-235 has been consumed and it also contains fissile isotopes of plutonium, which can be used in fast-breeder reactors. Since about 15 metric tons of irradiated fuel from an advanced gas reactor can be recycled once in a thermal reactor to produce as much electricity as 250,000 metric tons of oil, recovering the unused uranium and plutonium is a very worthwhile exercise.

At the same time a whole host of other products of the fission reaction are present in the fuel and these have to be removed before any of the useful material can be reprocessed. This is literally nuclear waste and in Britain the process begins at Sellafield, where spent nuclear fuel is recycled.

When the fuel emerges from the core of the reactor it is very hot and highly radioactive and so has to be allowed to cool down. This takes place first in a deep, water-filled pond at the power station, and again at Sellafield when it arrives. Everything is radioactive because the structure of the atoms has been changed by the continual bombardment of neutrons inside the

The examination of irradiated nuclear fuel elements carried out by remote handling in large, thick-walled concrete shielded caves

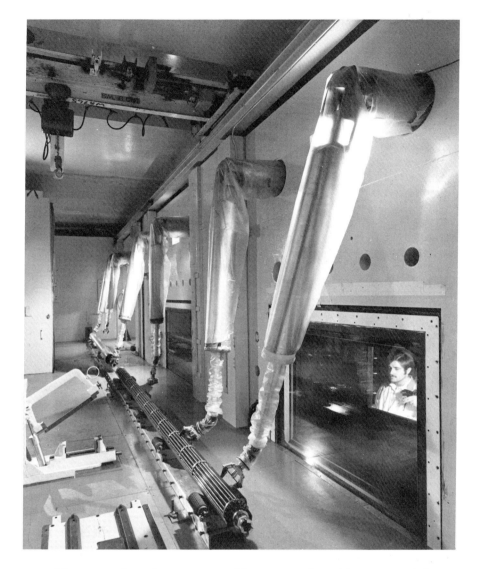

core. The atoms have become unstable as a result and have turned themselves into isotopes giving off radiation. However, the radiation is not the same for each isotope. Different isotopes emit different types of radiation.

## Radiation
The first of these are the alpha particles. These are streams of small particles, each consisting of two protons and two neutrons, and they have very little penetrating power; a sheet of paper or human skin will stop them. They are not dangerous unless they get into the body by breathing or swallowing, or through a break in the skin. Next there are beta particles, made up of streams of free electrons, and gamma particles, which are similar to X-rays but more penetrating. Together with a small quantity of neutrons which are also produced, beta and gamma particles can cause damage when entering or passing through the body. They need denser and stronger barriers to stop them; and both gammas and neutrons can only be screened by thick layers of concrete or water.

However, the type of radiation is not the only important factor. For although an isotope's radiation is affected by chemical changes, temperature or pressure, its intensity does decay with time. Each material is characterized by its 'half-life', which is the time it takes for half the radioactivity to decay, and this can vary from fractions of a second to billions of years. In general the most radioactive materials, emitting intense penetrating radiation and requiring heavy shielding, decay to negligible levels quite quickly. Longer-lived radioactive materials tend to emit much lower levels of radiation.

The radioactive substances in a nuclear reactor can be split up into three main groups, according to how they are produced. The first are the fission products, a wide assortment of new atoms formed out of the split fragments of uranium nuclei. They are generally beta and gamma emitters and many have short half-lives: only a few have half-lives longer than a year. The second group are heavy atoms called the actinides. These are formed when the uranium atoms absorb some of the neutrons produced in the fission process, but which do not create a reaction. Instead new elements are formed, the most important of which is plutonium-239. Although they are mostly alpha emitters, they do have very long half-lives. The half-life of plutonium-239, for example, is 24,000 years. The third and last group is

Advanced Gas Reactor fuel elements in a cooling pond at one of Britain's nuclear power stations

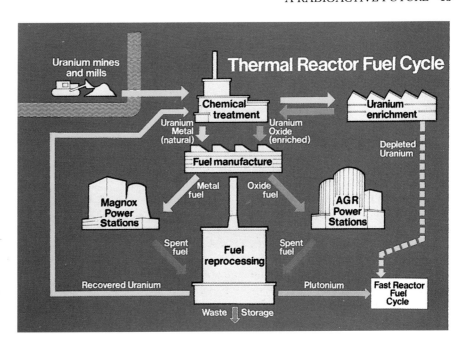

**Thermal Reactor Fuel Cycle**

RIGHT Diagram showing the complete nuclear fuel cycle
BELOW An aerial view of the Sellafield reprocessing centre in Cumbria

less homogeneous than the others, consisting of all the substances produced by various secondary processes. One of the most hazardous is carbon-14, a beta emitter with a half-life of 5800 years, better known for its work in helping archaeologists and others trace the age of ancient relics.

But how radioactive is this waste? If compared to the natural level of radiation coming from the original uranium ore out of which the fuel was made, the used fuel rods would be several million times more radioactive. After spending a year in cooling ponds they would be something like 10,000 times as active as the ore, and only after about 1000 years would the radiation have dropped to a figure approaching that of natural uranium. With levels that high, the problem of containing this radiation so that it has no harmful effects on the environment or on human beings has been exercising scientists since the 1950s. In fact it is only just recently that serious long-term solutions have been suggested.

**Highly active radioactive waste**
Disposing of highly active waste is the ultimate phase of the nuclear cycle. It has to be contained and isolated in such a way that any leak or return of the radioactivity into the soil, water or air should not only be almost impossible, but also should not present a risk to us or the environment in the future. There have been a number of proposed solutions, some of which seem to belong more to the realms of fantasy than reality. There is, for example, the suggestion that the waste should be sent up by rocket into space, aimed at the sun. However, the most recent solutions have involved burying it deep in the earth after turning it into either solid glass blocks or synthetic rock.

After the re-usable uranium and plutonium have been separated from the spent nuclear fuel by immersing it in a bath of nitric acid, the solution that is left is made up of a cocktail of radioactive ingredients. The mixture is distilled further into a more concentrated form and then stored underground in double-walled, stainless-steel tanks at Sellafield where it undergoes constant stirring, cooling and monitoring. Each tank, containing large amounts of radioactivity, has more than a kilometre, or three-quarters of a mile, of cooling pipes and is surrounded by a 2 m (6½ ft) thick concrete enclosure. According to British Nuclear Fuels, who operate Sellafield, the waste 'would be perfectly safe ... as a liquid in these tanks for many decades'. There are 16 such tanks in use at the moment, amounting to just over 909,000 litres (240,000 US gal) of waste and it represents the amount of highly active liquid waste from the last 30 years of nuclear power. Together with a small amount stored at Dounreay in Scotland, the total waste produced by reactors currently operating, under construction, or authorized by the year 2000 would fill about 30 such tanks.

However, the tanks do not provide a long-term solution to the storage of this high-level waste. Even stainless steel will begin to leak after about 50 years. Mild steel tanks leak even quicker. At the Hanford plant in the United States leaks developed only 20 years after the waste was stored. For this reason, and others to do with ease of handling, it was decided to immobilize Sellafield's high-level waste into a material that required a minimum of human supervision, and would remain safe without any continuing care or maintenance by future generations. And at a plant that is expected to be in operation by the late 1980s, the high-level radioactive waste will be turned into solid borosilicate glass blocks.

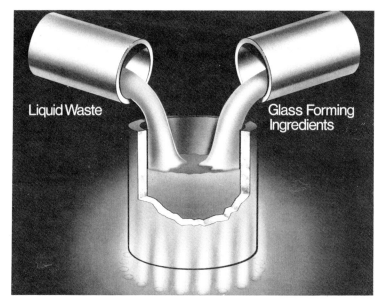

Liquid Waste

Glass Forming
Ingredients

ABOVE LEFT AND RIGHT The
vitrification process and the
result

### Radioactive glass

The technique, known as vitrification, has been under scrutiny for many
years. Research was started in Britain and France in the late 1950s and both
countries produced rival designs. The French process was put into operation at a plant in Normandy, but the British design has yet to receive any
approval. Instead Sellafield will use the French system. This will take
waste which has been in the tanks for at least ten years and finally place it in
a form from which it cannot spill, leak, vaporize or spread.

But how is it done? The first stage in the technique is to turn the highly
active liquid waste into a dry powder. By placing the waste in a rotary
calciner, it is evaporated and converted into brown granules. Now it is
ready to be mixed with crushed borosilicate glass so that it dissolves its way
right through the glass. In this form it does not matter if the glass is broken,
or if it is covered by water, as the waste cannot escape. The mixture is
heated in a stainless-steel furnace to 1000°C so that both the waste and the
glass are melted. This is eventually poured into a storage drum. The only
way now for the waste to escape is for the glass to dissolve. Since naturally
made glass can survive as long as 50 million years, it could prove to be an
ideal way of storing radioactive waste.

In Britain the general plan is to cast these waste glasses into solid
columns, each about the height of a room and the diameter of a barrel. A
year's waste from one 1000 megawatt nuclear station is expected to amount
to about seven of these cylinders. Each cylinder would then be sealed in a
stainless alloy container, chosen to resist corrosion for thousands of years
in a dry environment. But just in case it might become exposed to corrosive
natural waters, this container would be covered with an outer layer of some
other substance such as copper or lead, known to be able to resist salt water
for more than a thousand years.

By the end of the century Britain is likely to have accumulated about
10,000 steel bottles, each filled with a radioactive glass ingot, and the
number will go on increasing year by year. These will be stored at the
surface for a period of at least 50 years because of the heat they will still emit

– each ingot will produce about 2.5 kilowatts – and for more of the shorter-lived radionuclides to decay.

However, not every scientist believes that glass is the best option. It has been argued that although we know a lot about ordinary glass, we do not really know what will happen to radioactive glass. It will be constantly bombarded with radiation, which could damage its atomic structure. Some scientists are worried that this damage could make the glass more susceptible to dissolving in water, allowing the waste to escape before the radioactivity had died down to a safe level.

Over recent years, therefore, a group in Australia has been testing something that they know will last for thousands of years, even when it is radioactive: rock. The Australian researchers have taken a lesson from nature, observing that some rocks, such as granite, naturally contain radioactive elements, now at a harmless level, and contain them very securely for hundreds of millions of years. Could the scientists take the elements commonly present in naturally occurring radioactive rock and create their own synthetic rock to house high-level waste?

Professor Ringwood and his colleagues at the Australian National University decided to try. They have taken oxides of aluminium, calcium, titanium, varium and zirconium, mixed them together with radioactive waste and placed them in a hydraulic press. The press simulates the kind of natural forces that rocks were subjected to when they were first made. The mixture is then fired in an oven which mimics the heat of the volcano that forged the natural rock, and at the end of the process you have man-made radioactive rock.

According to Professor Ringwood, the synthetic rock has a number of advantages. Firstly, it dissolves less than glass when it comes into contact

LEFT AND OPPOSITE ABOVE LEFT
Designs for the burial of
nuclear waste underground

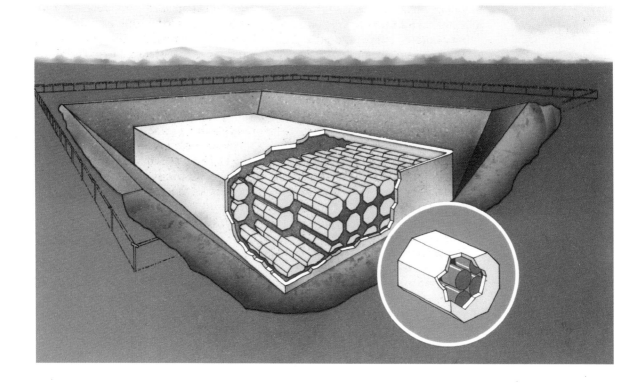

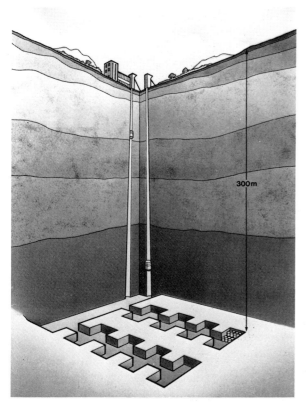 

ABOVE RIGHT The Drigg
National Disposal Centre for
low-level waste, near
Sellafield, Cumbria

with ground water and so is less prone to leak its radioactive contents. Secondly, glass can only accept 10 per cent high-level waste because the internal temperature of the stored glass blocks has to be kept to about 100°C: any hotter than this and the glass is liable to break up if it comes into contact with ground water. However, synthetic rock is so resistant to ground water attack that the Australian scientists are confident that the rock could contain up to 20 per cent of high-level wastes and need only be stored on the surface for 5–10 years, unlike the 40–50 years recommended for the glass ingots.

In theory this seems an ideal solution. A long period of testing will have to take place and a manufacturing process be developed before it can rival the glass technique. But if it proves successful, it could mean that all of the high-level radioactive waste that nuclear power produces could be embedded in rock in some way or another, because there are also proposals to store the waste underground.

## Buried underground

Whatever method is chosen to immobilize high-level waste, a long-term solution has to be found about its eventual disposal. The most favoured solution so far is to isolate it in geologically stable rock, in deep-mined depositories or boreholes well below ground. Permanently immersed in vaults, the rock would provide a more than adequate shield against radiation, and would absorb and disperse the heat produced by the decay of radioactivity and decrease the likelihood of any accidental intrusion into the store.

At the moment all work is aimed at proving that such an underground facility could be built, operated and closed safely using our existing mining and engineering experience. This is likely to mean the designing and building of one or more experimental facilities, with the results available in the early 1990s. The favoured grounds are granite, which is generally impervious to water, salt beds and clay-rich rocks.

In the old salt mines of Asse, near Hanover in West Germany, the sound of trucks reverberates through its passages. These are not ghosts of former days when salt was still mined. These are trucks carrying radioactive waste ready to be stored underground. For old mines such as Asse are being prepared to house intermediate or medium-level nuclear waste. This is not the highly radioactive kind, but it will still remain potentially dangerous for hundreds of years.

At the moment this kind of waste, which is created when a material comes into contact with the reactor fuel, is accumulating in concrete vaults and silos at nuclear installations. (It consists of such items as the outer packaging for the fuel assemblies, for example.) Another alternative store has to be found and in Britain there are several possible sites under investigation where the waste can be buried.

However, not all this waste that is being buried is a potentially dangerous as that at Asse. Low-level waste is already being buried in a trench at Drigg in Cumbria, not far from the Sellafield works and also where no one can reach it easily or see it; at sea.

**Dumping at sea**
Britain has been dumping low-level nuclear waste in the sea since the late 1940s. In 1982 it was the world's largest dumper, dropping around 400 tons of waste in concrete containers into water 6.5 km (4 miles) deep 800 km (500 miles) off Land's End. In 1983 plans to continue with the annual disposal provoked a storm of protest in Spain and Portugal, the nearest countries to the site; and despite growing international opposition, the practice was abandoned only as a result of a refusal by the seamen's trade union to handle the waste. The British government has now agreed to suspend all dumping at sea until a scientific review of its safety has been completed.

The theory behind the dumping of nuclear waste at sea is that the radioactivity should escape from its concrete containers gradually over a period of 10-20 years, as the containers are not expected to last any longer. This slow leakage of radioactive waste is then dissipated safely in a large body of water. However, a story run recently by the BBC's External Services Unit pointed out some rather worrying features.

This showed that not very much is known about the behaviour of the ocean at the enormous 6.5 km (4 mile) depths at which the containers are sitting. Recent research indicates that the water at these depths does not circulate to the surface as quickly as was previously supposed. Some experts have claimed that this makes deep-sea dumping safer still, as it would prevent the waste from reaching the surface. But other scientists are concerned that powerful ocean-floor currents, which have only recently been discovered, could carry contaminated water over long distances in a relatively concentrated form.

Another area of concern is the nature of the waste that is being dumped. Radioactive waste is divided into three categories - low, medium and high –

Low-level waste being
loaded aboard a ship at
Sharpness docks,
Gloucestershire, before
going out to be dumped in
the Atlantic

and this classification is made according to degrees of containment that
each type of waste requires. Some plutonium waste is defined as low-level
because the alpha radiation it produces is not very penetrating. Nevertheless, the radioactivity produced by plutonium is very intense. If ingested, or
if it comes in contact with skin, the smallest amounts of plutonium can be
deadly. Recently, the amount of plutonium in the waste being disposed of
at sea has been increasing and some people are worried that it could find its
way into the food chain through fish; and as it is very long-lived, plutonium
poses a continuing health hazard.

Interest has also been focused on the real level of alpha discharges. Some
researchers have pointed out that, until recently, discharges of plutonium-
241 were not taken into account in the alpha waste category. This may be
because plutonium-241, as distinct from other plutonium isotopes, was a
beta rather than an alpha emitter. However, plutonium-241, with a half-life
of just under 14 years, disintegrates into americium-241, which is both an
alpha emitter and of much higher radioactivity than plutonium-239.

Could it be that much greater amounts of alpha radiation have in fact
been expelled into the sea than was thought? Although most evidence
seems to suggest that there is no significant danger from the dumping that
has been carried on, the concern is that these operations may be leading up
to the disposal of intermediate-level waste at sea. Until the research is
available no further drums will be packaged ready for the ocean floor.

However, that is not the only route through which low-level waste is placed in the sea. British Nuclear Fuel's plant at Sellafield regularly discharges radioactive liquid effluent into the Irish Sea through a pipeline. The pipeline has been the subject of a great deal of controversy over the years, partly because, through accidents, higher levels of waste have been pumped into the sea than should have been. In November 1983, for example, highly radioactive waste was drained off together with low-level solvent, with the result that the beaches in the surrounding area had to be closed to the public for a number of weeks.

Changes in the procedures for waste disposal at Sellafield have resulted from this recent leak. But evaluating the safety of such sites and of using the sea as our radioactive dump is not easy. Part of the reason lies in the history of the Sellafield plant. There was a fire there in October 1957, when it was known as Windscale, after which radioactivity was released. It took years before the British public came to know that as much as 20,000 curies of iodine-131 had escaped into the atmosphere. A report on the fire by the National Radiological Protection Board (NRPB) also brought the release of as much as 240 curies of polonium-210 back into public attention. The NRPB suggests that the polonium may have killed up to 20 people.

Public anxiety about nuclear waste and its disposal has not gone away over the years, nor should it. The act of dumping radioactivity into the environment is a deed that is beset by dangers. However, the issue that has yet to be resolved satisfactorily is whether the dangers are significant or not to the health of the people in the vicinity and to the atmosphere. If they are, what is the level of tolerance? How much can the sea, soil or air take before damage is done? The question of public safety must be paramount and unless the issue is resolved, plants like Sellafield will always be looked at with a certain amount of suspicion.

The industry is in a difficult position. Satisfying the public that its activities are not dangerous has become almost an impossible task. The 'own goals' that the industry has scored has only made its case harder to accept. British Nuclear Fuels announced recently, for example, that 'a major programme of refurbishment and replacement . . . has taken place and as a result the amounts of radioactivity routinely discharged represent a dramatic reduction on the peak levels of the 1970s'. This is all very well, but it does raise concern for all those years in the 1970s when the pipeline pumped out what was then regarded as an acceptably safe amount of waste. If these levels are no longer acceptable, do we really understand fully what the safe levels are?

Public opinion accepts that in any industry accidents regrettably can and do happen. Industrial diseases have been identified and treatments found, and the risk of accidents is being dramatically reduced. The difference with the nuclear industry is one of scale. Not only are workers affected, but it provides an intangible threat to the whole population. Radiation cannot be seen or felt and its effects may not appear until years after contamination. The whole issue of safe nuclear power rests as much with the legacy of the waste as it does with the safety of reactors. The British Government White Paper in 1982 on high-level waste disposal stated that it was 'leaving the decision on disposal for a future generation'. In its view, 'we in the present generation have a clear moral duty to formulate the options [for disposal] as we see them at present, and to develop the supporting scientific and technical knowledge, so that they [i.e. future generations] will be better placed than we are to make the eventual choice.'

So does the question come down in the end to ethics and not technology? Is it that the morally correct course would be to avoid creating more waste until we are sure that an option which is practical, safe and publicly acceptable exists? Are we, by making use of nuclear power, imposing an effectively permanent burden on our descendants, one that will require constant vigilance and expertise well into the future?

But, would we be justified in ignoring nuclear energy and confining ourselves to fossil fuels, thereby depleting them for future generations? What will provide reliable, economic base-load electricity instead?

Whatever your point of view, it is clear that the present nuclear power programme will leave a legacy that will outlive its own technology. But is that technology safe in the first place, or could a serious accident happen?

### Three Mile Island

There have already been major reactor accidents throughout the world, and most varieties of reactor design have run into serious problems. A whole host of statistical evidence based on risk analysis has been put forward showing that the risk of a nuclear accident is a very remote one. Thankfully it is, and the nuclear industry goes to great lengths to try to ensure that no accidents happen. Yet scientists' perceptions of the chances of something unfortunate happening vary. One study in the United States based on all the mishaps at nuclear plants from 1969 to 1979 concluded that the likelihood of a major accident was around one in a thousand years of reactor operation.

This contrasted strongly with the usual assessment of one accident in 20,000 years of operation. The study, carried out by the Oak Ridge National Laboratories for the United States Nuclear Regulatory Commission, represented a sharp reassessment of the risks of nuclear power. With 72 reactors operating in the United States, it would suggest that an accident like that at Three Mile Island could have been expected every 10–15 years, given the equipment in place at the plant in that period.

What happened at Three Mile Island shook the confidence, and in some cases the credibility, of some of the industry's most forthright proponents. On the one hand, it terrified the local population around the site and unnerved the public around the world concerning the safety of PWRs. On the other, it removed some fears, as it showed that far from a catastrophe, one of the worst accidents that could conceivably happen caused no injuries or deaths.

The Three Mile Island reactor at Harrisburg, Pennsylvania, has been described as the 'largest tombstone in America'. For the events that took place in the control room on 28 March 1979 have, as it were, laid a wreath at the nuclear industry's door. No new reactors in the United States have been ordered since then and Three Mile Island's operators have been left with a bill for $1000 million to clear up and decontaminate the building, work that will continue for the rest of the 1980s.

A faulty valve and a series of human errors led to large-scale melting of the zirconium alloy fuel rods. Although only small amounts of radioactive gas were released into the surrounding countryside, the reactor core was devastated. When three years after the accident, the sealed-off reactor building was opened and volunteers entered the building, the extent of the damage was revealed. In the intense heat of the accident 90 per cent of the fuel rods had burned, about half of all the zirconium in the reactor had oxidized, and up to 70 per cent of the zirconium cladding had become so embrittled that it had lost its structural integrity.

The aftermath of Three Mile Island reverberated around the world. The incident may have frightened many, but it also had the effect of galvanizing the American regulatory authorities into action. The regulations concerning the construction of nuclear power stations and their safety procedures tightened. The history of the Diablo Canyon plant on the Californian coast is perhaps a good example of where the consequent increased vigilance uncovered serious shortcomings. Started in 1978, it had to be completely redesigned and strengthened four years later to take account of an offshore earthquake fault. Its start-up has since been delayed because of 318 different reports of shoddy workmanship and difficulties at the plant which had to be investigated by the United States Nuclear Regulatory Commission (NRC).

The new ground rules have not just made an impact on the safety of nuclear reactors; they have also had a profound effect on their financial viability. The tighter standards have meant spending a great deal more money on construction, which often has had to be repeated to meet the requirements of the NRC. Nearly everywhere in the nuclear power plant industry in the United States, enormous cost overruns have been commonplace. In Louisiana one electricity company is experiencing an elevenfold increase in the price of its reactor, estimated to cost only $230 million a decade ago. In South Texas two 1250 megawatt units will cost nearly eight times more than the originally planned $1 billion.

As a result of these overruns, the economics of nuclear power in the United States has now begun to be examined very closely. Large numbers of stations have been abandoned, some even during construction. In Britain, the Sizewell inquiry will examine the history of overruns in our own nuclear power industry with the Advanced Gas Reactor. However, the station proposed for Sizewell is of the Pressurized Water Reactor type, the same design as at Three Mile Island. The station, if it is given the go-ahead, will have a number of major modifications to ensure that some of the PWR's faults are not reproduced in England.

### Failsafe reactors?

The controversy about the safety of the PWR design and its economic viability still continues. Obviously, all nuclear plants will be made as safe as they possibly can be. So the development of two reactors which have been called 'intrinsically safe' has been attracting a great deal of interest throughout the nuclear industry. They are stations whose safety depends not on mechanical or human intervention but on the workings of well-established principles of physics and chemistry. Human fallibility and the reliability of failure-prone equipment would no longer be involved.

There are two designs: the Swedish Process Inherent Ultimately Safe Reactor (PIUS) and the Modular High Temperature Gas Reactor (HTGR). The PIUS is a PWR immersed in a pool of borated water, and the pool itself is inside a prestressed concrete vessel that operates at a pressure of 100 atmospheres. The circulating coolant is separated from the water in the pool and this is where its safety comes from. Any loss of coolant, as in a leak or crack, automatically causes the borated water to flood the reactor, drowning the chain reaction and holding the situation stable for seven days, totally unattended.

Three Mile Island Power
Station – Harrisburg,
Pennsylvania, 18 March
1979

The HTGR works on a different principle. It is a graphite moderated reactor but its safety lies in the fact that it operates at such low power compared with normal nuclear stations (100 megawatts), and at such a lower power density, that even a total loss of coolant would not cause the temperature of the fuel to rise to a dangerous level.

The next few years will see whether a serious effort will be launched to explore the possibilities of these new designs. PIUS was created with the intention of removing any reasonable doubt in the mind of the 'enlightened layman' about the safety of nuclear reactors. It remains to be seen whether or not it will be constructed.

The health of future generations depends upon the safety of nuclear power and its related technologies. After 40 years of research and development one question still remains open: are our energy supplies worth the price of eternal vigilance and a high level of safety and responsibility? On the other hand, since we know that fossil fuels have a finite life, is there any serious alternative to nuclear power? The rest of this book is an attempt to find out.

# CHAPTER FOUR
# A Ray of Sunshine

In less than 20 years there has been nothing short of a revolution taking place in the world of solar power. Then the idea of building huge solar power stations, able to capture the immense amounts of energy generated by the sun, belonged to the world of science fiction, a wild boffin's dream that might emerge some time in the twenty-first century. But now solar power is on the brink of commercial reality, and in some parts of the world its use is already economical. All over the world, from the Soviet Union to the Californian desert, money has been poured in to explore the various ways of harnessing the sun's energy; the solar industry has become big business.

But why this sudden explosion and how has it come about? Does solar power provide the long-awaited answer to the world's energy problems? The sun certainly could meet the world's demand for energy many times over. The amount of solar radiation that reaches the earth's atmosphere in one hour could meet the world's annual energy needs if we could exploit it fully. Even in the British Isles the amount of solar energy received is 80 times more than what we need at present from primary energy sources like coal and oil. And what is more, the sun delivers sunshine free to all people. It does not pollute the atmosphere and it is a secure source of energy that will not run out. It has been estimated that it will keep on burning five million metric tons of hydrogen a second for at least another five billion years.

OPPOSITE Sunrise
BELOW The Solar Chimney project at Manzanares, Spain (see page 57)

So how does it work? The sun is a giant nuclear fusion reactor and it gets its energy from the enormous atomic furnace at its centre, where temperatures could be as high as 15 million degrees C. At that temperature atoms are moving so fast that nuclear reactions take place as they bump into each other, converting hydrogen into helium. The energy that is generated then works its way through the 772,500 km (480,000 miles) between the sun's core and its outer edges (the photosphere). The process takes so long that today's 'sunshine', which took only just over 8 minutes to travel the 148 million km (92 million miles) to earth, was generated in about 8000 BC.

The sun's energy, in the form of electromagnetic radiation, emerges into space at a rate of $4 \times 10^{26}$ watts. However, the earth only intercepts about one-billionth of this massive output and its atmosphere modifies what is received even more. A spacecraft orbiting the earth, just outside the atmosphere, would receive something like 1350 watts on a collector of 1 m$^2$ (10.8 sq ft) which was facing the sun, but the earth's surface receives much less than this. When the sun's energy reaches the atmosphere about 30 per cent is reflected or scattered straight back into space. Air molecules, naturally occurring gases, water vapour and clouds reduce the amount still further, although it is not lost energy. It forms the 'power station' of the earth's climate, driving the wind and water cycles that regulate our environment and, in turn, provide further potential energy sources.

How much solar energy then is left for us to collect? It has been calculated that at noon on a clear summer's day in London, the intensity of solar energy falling on 1 m$^2$ (10.8 sq ft) of horizontal earth could reach a maximum of 1000 watts. But introduce a cloud or two and that figure can fall down to about 200 watts per square metre.

In the British Isles we have a further problem. We receive about 50 per cent less solar energy than, for example, California or Israel, and about half of our sunlight is diffuse light. This means that the sunlight has been scattered or reflected by the atmosphere or ground and cannot be focused; we cannot therefore utilize solar furnaces and devices that employ mirrors to focus or concentrate the sun's rays on an economic scale, as they need direct light to work. But although our high proportion of cloudy days may make one approach difficult, there are other ways of exploiting solar power. They range from high technology solar cells which convert sunlight to electricity to less sophisticated designs for providing hot water in our homes.

## Solar collectors

The general view of much solar technology is that it is untried, untested, experimental and uneconomic: a resource that is slowly being developed and researched for the distant future. This is quite true in certain respects, but some solar technology is nearly a century old. In 1897, 30 per cent of the houses in Pasadena, California, were equipped with solar water heaters; and the original flat plate collector, a glass-covered box with water tanks inside, was patented in Baltimore, Maryland, in 1891. Florida, too, had its fair share with over 60,000 of them in 1940. There are now estimated to be well over four million in use, and the world's solar collectors in the early 1980s yielded about 0.01 per cent of its annual energy consumption.

With figures like this, solar collectors obviously are not going to solve the energy crisis overnight. The market and the technology, however, are still growing, and growing fast. Japan, which has over three million collectors on its roofs, already plans to equip 20 per cent of its housing stock with them by 1990, hoping to save the country the equivalent of five million metric tons of oil a year. In Israel, nearly half the houses are equipped with a solar water heater, and by 1985 the Israelis hope to have reached over 60 per cent. The Canadians, too, are aiming to collect 12 per cent of their energy requirements by solar means by the year 2000 and the Australians are no less ambitious.

There are basically two ways in which the sun's energy can be utilized. One is to use the heat directly to warm water or to drive machinery like turbines; the other is to produce an electric current from the sunlight by means of a solar cell. Solar collectors are devices that trap the sun's heat and they come in a wide variety of shapes and sizes. There are very large installations which create intense heat directly for industrial processes, or boil water or some other liquid to produce electricity via a generator and turbine. These systems are capable of creating temperatures up to 900°C. Alternatively, there are the much smaller solar panels which can be used on the roof of any ordinary house.

### The flat plate collector

Take a shallow wooden box, fill the base with high temperature insulation material, place a blackened central heating radiator on top, cover the whole lot with glass, and with a little bit of care and luck you will have built yourself a makeshift flat plate collector. Obviously this particular model will not be necessarily the best solar collector on the market, but it does illustrate the simplicity of the design. The basic solar unit has a metal plate or absorber to collect the heat and a pane of glass to trap it. Attached to the plate are pipes with a fluid flowing through them. This fluid, which is normally water or some kind of anti-freeze, is heated up by the sun and then pumped through the pipes to a storage area, often via a heat exchanger. It then transfers its heat into either a hot water or central heating system

Between 20,000 and 25,000 flat plate collector systems have been installed in Britain, all of them supplementing some other form of heating and saving on fuel bills. A typical system, with a collector area of about 5 m² (54 sq ft), should provide a family of four with all the hot water they need on many days during the summer. In the winter existing systems, like gas and electricity, need to be used to reach the required temperatures and any solar installation has to be seen as a long-term investment.

### Just a lot of hot air?

Another alternative to the flat plate water collector is one that uses air not a liquid to carry the heat. They are cheaper and easier to make and operate on the same principles, with air drawn through the collector and into the house via ducts. In Britain one experimental system built by the Building Research Establishment in Watford, Hertfordshire, covers the entire south roof slope of a solar house, and has reached temperatures of 55°C on a hot day. Heat pumps are used to transfer the heat from the air stream to the hot water tank as well as helping to heat living areas in winter.

Small solar air collectors may prove interesting, but one scheme put forward by Professor Schlaich of Stuttgart University in West Germany shows that solar power could, literally, end up as a lot of hot air. He has designed a solar chimney and the first prototype was completed in Manzanares in Spain in 1982. It has a 200 m (656 ft) high, 10 m (33 ft) wide central tower made of corrugated steel, which is surrounded by a solar collector area in the form of a transparent plastic greenhouse. Just as in a normal greenhouse the air inside is heated by the sun, but Professor Schlaich has gone one stage further. Inside the tower he has linked the idea of hot air to wind power.

A turbine has been placed in the bottom of the tower, consisting of four 5 m (16 ft) variable pitch blades; and so as the air is heated in the greenhouse, it rises up the chimney, turning the turbine in the process. Fresh air is sucked in from outside the collection area to keep the cycle going.

The advantage of the chimney design is that it does not require direct solar radiation, but can still generate electricity when the sky is overcast and the radiation is 100 per cent diffuse. Professor Schlaich and his colleagues have calculated that a solar chimney that could generate 50 megawatts or more can be economic. Whether or not they will be built is another matter.

### Solar ponds

The best way of exploiting the sun's energy might be through salt water contained in a man-made solar pond. For unlike all the efforts scientists have been making in the Third World to take the salt out of salt water, researchers in this field have been trying to increase the concentration of salt in the water to get power out of the sun. The idea of the solar pond originated when it was found that some lakes with a high natural salt content developed very high temperatures well below the surface. In Hungary, for example, Lake Medeve reaches temperatures of 71°C at depths of about 1.2 m (4 ft) in late summer. Scientists found that the deeper you go, the hotter and saltier it can get. Natural solar ponds work because the sun's rays pass through the transparent upper layers of water and are absorbed by

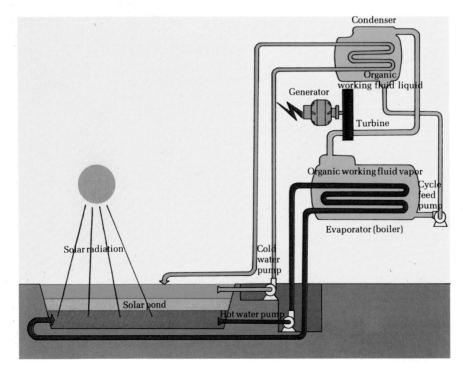

A solar pond

RIGHT The solar furnace at
Odeillo in the French
Pyrenees
BELOW One of the latest
parabolic collector plants to
be built. This one is near
Warner Springs, California

the much denser, saltier layers at the bottom. So why not copy nature and build ponds where the salt concentrations are carefully graduated to keep the water as hot as possible?

In a solar pond the salt layers trap the sun's heat at the bottom, and to prevent these layers from evaporating an insulating layer of fresh (i.e. saltless) water is placed on top. This prevents the heat from travelling back up to the surface, which is thankfully just what normally happens in ordinary sea water: otherwise going for a swim would be a very unpleasant experience indeed.

The hot salt water is then extracted from the bottom of the pond by pipes. Now since at 90°C the water is below boiling point, it obviously cannot be used to drive a steam turbine. Instead, researchers have found that by using the water to heat a chemical with a much lower boiling temperature: usually a refrigerant like freon. The vapour from the refrigerant is then used to generate electricity.

Solar ponds have aroused a great deal of interest and are being studied in Israel, Australia and the United States. In Israel one pilot plant has been opened on the shores of the Dead Sea. The 7000 m² (8372 sq yd) pond, believed to be the largest working solar collector in the world, is operated by just one man and is aimed at proving that they are technically and economically feasible propositions. Even larger schemes are being planned, and the Israelis hope to create huge solar lakes on the Dead Sea. The United States Jet Propulsion Laboratory seems no less ambitious. It hopes to be able to transform the Salton Sea, a salt lake of about 930 km² (360 sq miles) in southern California, into a 5 megawatt solar pond plant. If the initial scheme looks successful the Salton Sea will be gradually converted over to producing a massive 600 megawatts.

Schemes like the Dead Sea and Salton Sea could turn out to provide reasonably cheap power, and since the technology is still developing most people are hopeful. They might stand an even better chance if recent developments in England at Sussex University can be utilized. These have shown that extracting the heat at the bottom layer of the pond is only part of what can be done. There is also a substantial amount of heat in the insulating layer. By extracting and using this heat it is possible to make the pond more efficient. This can be done by immersing water pipes at different levels of the pond, providing hot water at various temperatures. The hot water can then be used for other purposes like heating houses or greenhouses.

**Solar power stations**

All of us have at some time spent too long worshipping the sun's rays on holiday: but if you pay a visit to a 7 hectare (17 acre) site high up in the western Pyrenees in France a sign at least warns approaching visitors of the perils that await them. It says, 'Danger! Solar Rays'. You can see why when you reach the top of the site as there are 200 mirrors arranged in a perfect semicircle, all leaning like some vast pagan monument in homage towards a tall 100 m (328 ft) concrete tower. It is in fact Europe's largest solar energy installation, generating up to 2.5 megawatts, and is one of the many solar power stations that have been constructed in the last ten years. There are several different systems in operation, of which more in a moment, but they all rely on focusing or concentrating the sun's rays on to a central point.

The idea of focusing sunlight to create heat and energy is about as old as man's earliest records. The Ancient Greeks used concave vessels of brass to light sacred flames in temples (this method is still used to light the Olympic flame). Leonardo da Vinci suggested that a giant parabolic mirror 6.4 km (4 miles) across could be built which would 'supply heat for any boiler in a dyeing factory', creating a pool of boiling water. Focused sunlight was even used to power machinery back in the nineteenth century; solar-powered engines were exhibited at Tours in France in 1874. A cone-shaped mirror 2.6 m (8½ ft) wide focused sunlight to generate enough steam to drive a ½ horsepower engine at 80 strokes a minute. So what are the designs of the 1980s?

There are two main types of concentrators, linear-focusing and point-focusing devices, and both track the sun throughout the day to maximize what sunlight there is. Linear-focusing collectors are long, silver troughs that use mirrors or lenses to focus light on to a pipe that runs the length of each collector. The pipe carries the heat transfer fluid, usually oil or pressurized water. The efficiency of the system depends upon its 'brain'. This is a small sensor that notes the position of the sun, and relays the information to a series of electric motors which keep the collectors correctly aligned. However, they only track the sun on a single axis, east to west. Point-focusing devices are more sophisticated. They move along two axes and can concentrate the sun's direct light on to one central absorber. One particular design is the parabolic dish collector.

## Parabolic collectors

Parabolic collectors tend to be very complex because the receiver is small in relation to the total area that is devoted to capturing the sun's energy. High levels of optical precision and guidance systems are needed, but most of the early problems have been overcome and these collectors have proved quite successful. The first big parabolic collector was built in the late 1960s, when the French opened an experimental 1 megawatt solar furnace at Odeillo in the Pyrenees. (The 2.5 megawatt station mentioned earlier is not far away.) It has 63 large computer-controlled heliostats spread over 4 hectares (10 acres) and all the sunlight is concentrated on to a receiver using a parabolic reflector that is 42 m (138 ft) wide. Temperatures can reach over 3000°C. Odeillo's intense heat is used to melt certain materials for industrial purposes, although its prime purpose is experimental. Heat from smaller and less intense systems is now applied in oil refineries, breweries and even potato-processing plants, particularly in the United States.

Other recent installations are producing electricity, including one at Nio in Japan, which uses a bank of nearly 2500 flat plate collectors to track the sun, and just over 120 parabolic trough reflectors to collect the solar radiation and further concentrate it. Power is produced in the conventional way using high-pressure steam in boilers, which is expanded through a turbine to drive a generator.

Not all solar power stations, however, are based on the parabolic trough design. Perhaps the favourite system is the one that uses a central receiver, or 'power tower' as it is called.

## Solar power towers

Central receivers work broadly on the same basic principles as parabolic trough collectors. They provide high temperature liquids, either steam or molten salt, to drive a powerplant producing electricity, and they use

thousands of sun-tracking mirrors to focus the sun's rays on to one receiver mounted on the top of a tower. Inside the tower the focused sunlight heats up the fluid inside a whole series of blackened pipes, which then carry the energy down to the ground. Most towers include reservoirs which enable the heat to be stored until it is needed, usually when the sun goes down or when there is bad weather.

ABOVE LEFT The Solar One power tower at Barstow, California
ABOVE RIGHT British solar cells in use on a recent expedition to Mt Everest

Solar power towers have been built all over the world as practical experiments to help in trying to understand the technology more fully. Early problems have partly concerned which heat transfer liquid to use. Molten salt has become attractive recently as the best liquid to use, because it retains its heat for much longer periods than steam and gets round the expensive difficulties involved in controlling high-pressure steam. However, one company has analysed the possibility of using helium, operating at over 800°C in a plan to generate 100 megawatts from one site.

The biggest system at the moment is that at Barstow, California, which first produced electricity in April 1982. It was not the first, however. That honour belonged to a small plant with 121 small mirrors, built by the University of Genoa in Italy in 1965. The Genoa device was a significant step forward and Europe has continued this trend by being the first to build a large-scale experimental solar power plant, at Adrano in Sicily. Feeding its 1 megawatt into the national electricity grid, the Sicilian project is part of a large EEC programme designed to build up our expertise in how to build and operate these huge arrays. There are other power towers in Spain, Japan, New Mexico and the Soviet Union.

One question that arises is how much land do these systems take up? The Solar One station at Barstow is situated in the Californian Mojave Desert, which has more than 300 cloudless days a year. But the site with its 1818 heliostats occupies about 40 hectares (100 acres) of land, and some critics of the solar power programme have suggested that deserts are the only place where they can be sited.

Solar One cost $142 million to build and such is the level of interest in the technology that most of the big aerospace and power companies are involved in such projects in the United States and Europe. However, given the high level of progress, is there not a risk that the technology will advance so fast that the Barstow plant and its counterparts will be condemned to rust away as monuments to an abandoned technology, rather as Stonehenge remains today as a memorial to some forgotten religion? One vital factor will be the high cost of the electricity produced; this will have to fall for solar power plants to become an economic feasibility. The next decade will prove crucial; but Solar One's operators are already planning to build a 100 megawatt plant because it has been so successful, and there are proposals that existing fossil fuel power plants could be 'repowered' using solar thermal energy, so the outlook looks promising.

### Solar repowering?

How could solar energy plants 'repower' an existing fossil fuel station? It has been suggested that if a solar power plant could be installed next to a conventional one, the solar heat produced could be used to fuel the normal steam generators. If no solar power was available, or it was not sufficient and needed topping up to reach the required temperature, the existing oil, coal or gas boilers could be brought into action.

If such an arrangement could work, this would significantly reduce the

'The Quiet Achiever', a solar vehicle that recently completed the 4800 km (3000 miles) from Perth to Sydney, Australia in 18 days

LEFT AND BELOW A solar car
and a solar cycle invented
by Alan Freeman. The solar
cycle is a solar/battery
hybrid and has a range of
58 km (36 miles) non stop at
a maximum speed of (24 km/h
(15 mph). The car, too, has a
top speed of 24 km/h

amount of fossil fuels we use. One survey taken in 16 southwestern states of the United States reckoned that about 13,000 megawatts of solar power could be accommodated in repowering schemes, saving about 11 per cent of the oil and gas currently being consumed by electricity companies. Repowering could also be used in industry to provide heat in the same way.

To see just how realistic the idea is the United States Department of Energy is now funding the conversion of four existing power stations. The biggest of these is in El Paso, where a solar power tower capable of producing up to half of the station's steam requirements is being built, and should be ready in 1987. Hopeful as these developments may seem, none of them is thought likely to place solar technology on a sound commercial footing or become widely established just yet. Most hopes are pinned on one so far unmentioned favourite: photovoltaic or solar cells.

**Photovoltaic cells**
A photovoltaic cell produces electricity from the second most common element on earth, silicon, the basic constituent of sand. It consumes no chemicals, has no moving parts, needs no maintenance and, providing you do not drop a brick on it, it should last for years. It is self-starting and self-sustaining – you do not even have to press a switch. A spectacular demonstration of the power these cells can produce was the flight of the *Solar Challenger* aircraft in 1981. With its wings covered in photovoltaic cells, this plane made the trip across the English Channel in style, even if it did take just over 5 hours and 20 minutes from Cormeilles-en-Vexin near Paris to Ramsgate in Kent. However, solar cells are not just featured in experiments like the *Solar Challenger*. They are used widely in a large number of applications: operating lights on navigation buoys at sea; in telecommunication devices in remote areas; in anti-corrosion systems to protect oil pipelines; to provide solar-powered cookers and refrigerators; and even to cross Australia in 20 days in a solar-powered car.

Impressive as designs like the *Solar Challenger* are, however, the flight did as much to demonstrate the problems of solar power from photovoltaic cells as it did its potential. The cells needed a constant supply of bright sunshine to ensure the propeller received the 2.5 kilowatts it needed to keep the plane aloft; and to collect that much power the *Solar Challenger* needed 16,128 photovoltaic cells worth about £148,000. That kind of money would buy enough aviation fuel to fly an ordinary small aircraft around the world four times. If solar cells are to make any impact at all they have to do much better than that, and all efforts are now focused on achieving cheaper and more efficient cells.

The principle of photovoltaic power – that certain substances produce electricity when they are exposed to light – has been known since the end of the nineteenth century. But the first practical solar cells had to wait until 1954 to appear, when scientists at Bell Laboratories in the United States found that single crystals of silicon could turn sunlight into electricity. What happens is that as light energy, in the form of photons, strikes silicon atoms, enough of their energy is transferred to knock electrons free from the atoms to produce an electric current.

Compared with power from conventional methods like hydro-electricity, coal or oil, however, early single-crystal silicon cells were prohibitively expensive at over $1000 per 'peak' watt (a 'peak' watt is the amount of power that is generated at noon on a clear day when the sun is

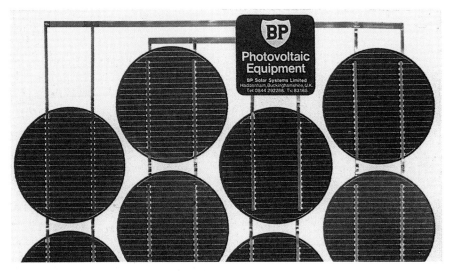

Part of a photovoltaic array

strongest). At prices like that the only place where the technology could be used was in space, where the ability to convert sunlight directly into a few hundred watts of electricity for extended periods, with cells that did not weigh too much, justified the high cost of photovoltaic power. Solar cells made their début on the United States satellite Vanguard in 1958, and have been used on thousands of satellites ever since.

But what are solar cells made of? A typical photovoltaic unit consists of

BELOW, OPPOSITE AND OVERLEAF The use of amorphous silicon is creating a great deal of interest and may form the basis of a new generation of very efficient solar cells. The cells are made in a fully automated, continuous process, using techniques not unlike those used to manufacture newsprint or photographic film. It can bend, is light and is very portable.

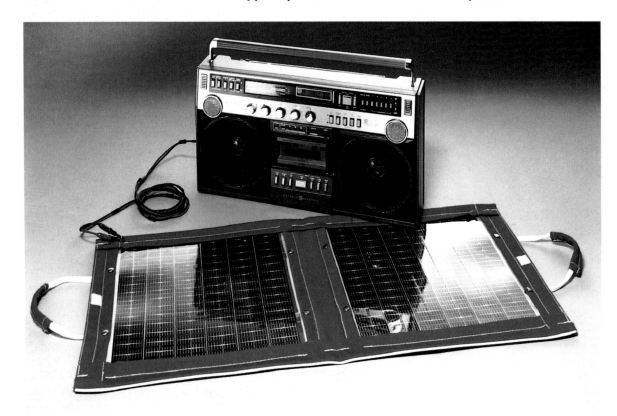

between 20 and 40 cells, with each cell made out of a circle of silicon about 50 mm (2 in) in diameter. The cell can be made out of a variety of semi-conducting materials, including gallium arsenide and cadmium sulphide, but silicon is by far the most developed and most popular. A number of these modules are connected in series together, and are usually linked to a battery which both stores the electricity and feeds it out at an even rate. The battery is a vital part, because the current produced can vary from moment to moment and will drop as clouds pass overhead.

The choice of material for making the solar cell has become the focus of much of the research work recently, as laboratories and companies try to find cells that convert sunlight into electricity more efficiently than before. The present generation of cells only converts up to one sixth of the sunlight received: with current technology it would take an area of 16 km$^2$ (6 sq miles) to replace a 2000 megawatt power station like the one at Didcot in Oxfordshire. However, different methods and materials are already emerging to reduce costs and increase efficiency.

The ultimate goal would be to find a material that could simply be sprayed on to a plastic sheet and rolled out at the end of a production line: and with the arrival of amorphous materials it looks as if manufacturers could do just that. The difference between amorphous silicon and the

normal crystalline version of the material now in use is rather similar to the difference between glass and rock. Glass is different from rock because it is cooled too fast for crystals to form. Instead of conforming to the usual tight discipline, its atoms are set in an amorphous pattern with no set structure. Amorphous cells are now produced using techniques developed for the printing industry, in which the silicon is laid down as thin film on sheets of stainless steel. At the moment the sheets are only 300 mm (12 in) wide and are produced at the rate of 300 mm a minute. But by 1988 it is hoped that the cells will be made on rolls 1.2 m (about 4 ft) wide and at a rate of 3 m (nearly 10 ft) a minute. The cells would then cost one-tenth of today's commercial crystalline cells, making photovoltaics competitive with other forms of electricity.

Another potential breakthrough has been found in Japan, where solar cell material can now be printed directly on to a glass base. This time the material is not silicon, but cadmium sulphide. It cuts the cost of solar cells by more than half and provides a conversion efficiency of around 13 per cent. Cadmium sulphide is also used in an experimental process devised by British scientists. It is laid down by electrophoresis (a method of separating chemicals on the basis of their molecular weight and electrical charge), and processing by lasers converts it into semi-conductor material. The work will not be completed until 1990 – and by that time who knows what will have emerged? Perhaps one possible front runner may be based on a compound called iridium phosphide, which is already well established in the electronics industry (ironically enough for converting electricity into light: it is the basis of most light-emitting diodes).

By 1990 we should also know the results of the massive programmes under way at the moment to test the practicality of photovoltaic power plants. The world's first large (1 megawatt) plant was completed at Lugo, near Victorville, California, in November 1982 and armed with the knowledge gained there, the company involved has started work on the first phase of a 16.5 megawatt plant at Carrina Plain, also in California. Not all work is concentrated in the United States. The EEC has been carrying out a major programme under which 15 photovoltaic plants have been built; the last of these was completed at Marchwood in Southampton in 1984. The plants, which are intended to show that photovoltaic power can be produced even in Europe's northern latitudes, are being used for a variety of purposes. One system is powering traffic control equipment at Nice airport, another is driving water pumps in Belgium, and a third is providing the entire electrical supply for the inhabitants of a small island off southern Italy. The largest plant producing 300 kilowatts, is at Pellworm the tiny North Sea island.

With a population of just over 1000 people, and nearly as many sheep, Pellworm 16 km (10 miles) off the coast of West Germany, is not the most immediate site you would think of for a pioneering project. The plant has more than 17,500 individual solar cells spread across an area equal to the size of two football pitches and provides power for the island's health spa. Any excess energy that is produced can be stored in a network of batteries for use at night or on cloudy days. Alternatively it can be fed back to the mainland and into the electrical grid. Just to make sure that the peaceful farming community is not disturbed by this high technology plant, the huge rectangular panels have been mounted off the ground, so that the island's sheep can go on grazing beneath them.

Ovonics photovoltaic
processor

### Photovoltaics – a viable alternative?

RIGHT AND BELOW The solar-
powered satellite – a science
fiction fantasy?
OPPOSITE The Marchwood
photovoltaic array,
Southampton

An EEC study projected, rather optimistically, that photovoltaics will be
providing up to 10 per cent of Europe's electricity demand by the year 2025.
Another major study, conducted by the Electric Power Research Institute in
the United States, showed that photovoltaics offered considerable promise
as a source of bulk electrical power by the turn of the century. The next
decade will prove crucial in deciding whether solar cells are going to be
technically and commercially viable for widespread use in Europe and the
United States.

Other parts of the world may not have to wait so long for solar power. In
some areas photovoltaics are already economical. One example is the use of
solar energy-operated water pumps. Over 100 photovoltaic units, power

Solar power when you want it, where you want it

pumping systems on wheels, are being used in villages and refugee camps in the Somali Republic, providing water for drinking and irrigation.

Since it is estimated that over 50 million of the world's poorest families exist on less than 50 million hectares (125 million acres), or less than 1 hectare per family, in the fertile, alluvial valleys, deltas and coastal plains of the Third World, a great deal can be done. Most of these areas experience lengthy dry seasons, during which cultivation proves impossible without proper irrigation. With adequate irrigation facilities beyond the financial means of small farmers, the development of small solar units capable of supplying water to farms where conventional methods are either inappropriate, or too expensive, might provide an answer.

In other situations photovoltaic systems are helping to fight disease. For the World Health Organization one of the biggest problems is getting any kind of immunization programme going, and then getting the vaccines to the patients. This is because unless they are kept cold, most vaccines decompose. A conventional portable refrigerator needs a reliable source of electricity, and in many places that just is not available. A photovoltaic-powered refrigerator that uses an array of solar cells has therefore been developed and could be used to save lives in the Third World.

For remote locations and for certain specialized uses, photovoltaics are already economical and are improving the quality of life for people who have never had access to electricity before. The key to the future of this technology lies in the application of mass production techniques to manufacture cheap solar cells. Its implications for the large electricity supply industries are also important and challenging. If photovoltaic plants become widespread, they would begin to challenge the structure of our present system as they would permit a greater devolution of power generation. One market research study went so far as to predict that by the year 2010 solar power would be providing 1000 times more energy than it does at present, with photovoltaics taking the lion's share at 88 per cent of generating capacity. If that turns out to be true then for many of us our power in tomorrow's world could come from solar cells on our roofs.

### Solar-powered satellites

High up in geosynchronous orbit, some 35,680 km (22,170 miles) above the earth's surface, 10 billion solar cells, mounted on an enormous aluminium girder structure and spreading over 50 km² (19 sq miles), gather the sun's rays and convert it into electricity. The direct current output is then converted into microwaves and beamed down to earth. Is this the ultimate dream of the solar scientists, nearer the realm of science fiction than of practical reality? For Peter Glaser, who suggested the idea in 1968, it is a deadly serious proposal; and for all its science fiction tinges, the concept of a solar-powered satellite (SPS) has its attractions.

Its main attraction is that it would be able, in theory, to capture all of the sun's radiation that reaches the earth. It would amount to about 30 per cent more than is available at noon on a clear summer's day, and there would be no interruptions from the weather or from the changes of day to night. The SPS would give a constant, 24-hour supply, and the power produced could be directed to anywhere in the world.

However, as might be expected, there are major difficulties connected with the project. Surprisingly enough, a study carried out by NASA and the Department of Energy showed that there were no major technical snags, but the SPS would only prove economical if photovoltaic cells came down in price and the parts could be successfully transported and operated in space. Although the former looks possible, the latter is totally dependent on the development of the US Space Shuttle. But it will not be shuttles that would be needed to build the SPS: rather Super Space Juggernauts, trans-porters capable of carrying much larger payloads in their cargo holds.

Other problems are the regular eclipses of the sun, which could cause temperature changes of up to 200°C; the accuracy of the microwave beams down to the reactors on earth, which would themselves cover an area of 80–100 km² (30–40 sq miles); and the considerable radio interference that the beams would cause, likely to make life for radio astronomers almost impossible.

For the SPS project to work, a whole series of satellites would have to be built, with each one capable of switching its load from one rectenna to another. One calculation showed that 60 satellites producing 5 gigawatts each could power the whole of the United States, and 40 the whole of western Europe. At the moment, however, the project is still firmly on the drawing board. According to the US National Academy of Science, solar satellites could become an interesting option some time in the twenty-first century. The European Space Agency has concluded that a fleet of SPSs could provide as much energy during the twenty-first century as fast breeder or fusion reactors.

Clearly the day of the solar-powered satellite is a long way off. However, one small step towards it was the launch of a huge solar wing from the Space Shuttle in 1984. Designed to power and prolong future space missions, it consisted of an array of 84 hinged panels that expand, like a concertina, from 100 mm (4 in) to a staggering 35 m (115 ft) across. The idea is that eight of these panels could power a 12-man space station. The first tests are rather more conservative, with only 900 solar cells (instead of the full 130,000) in place on the wing. Work at British Aerospace on the space telescope project, due to be launched in 1985, is also encouraging. There an array has been built covering some 33 m² (355 sq ft) with over 48,760 cells on it. It is all rather a long way from the pioneering days of Vanguard.

CHAPTER FIVE
# Power from Plants

Look through past issues of *Tomorrow's World* or certain scientific journals
and you will no doubt find potatoes powering clocks or lemons bringing
small light bulbs to life: odd items that are sufficiently strange to catch your
eye, but hardly enough to bring the battery industry to its knees. However,
plants and animals can and do produce energy. And this is far more
important and relevant to the world's needs than using vegetables to power
your television.

Plants use solar energy to harness the photons in the sun's radiation to
produce the energy they need, rather like photovoltaic cells do. The process
is called photosynthesis and it provides a vital link in the chain that
sustains life on earth. It regulates the health of the atmosphere, recycling
through plants all the carbon dioxide present on earth every 300 years, and
all the oxygen every 2000 years. It provides the oxygen we breathe. Photosynthesis works by trapping the energy contained in sunlight and using it to
convert carbon dioxide and water into energy-rich substances called carbohydrates. Plants use carbohydrates for the energy they need to grow, and
the same process is common to all animals and human beings as well. We
all need carbohydrates to survive.

The past results of photosynthesis have created our present sources of
coal, oil, gas and firewood – not to mention our food, fibre and chemicals.
The process still stores as much energy as the world needs, despite the fact
that on a global scale it has an efficiency of only about 1 per cent. But when
you have the whole earth to utilize, 1 per cent is quite enough. What

Trees in the fast lane. One
answer may be the giant
ipilipil

scientists are now trying to discover is how this biological conversion of solar energy can be harnessed as a source of energy for the future.

At a basic level, biomass-derived energy can be as simple as burning wood, either as logs or charcoal, for heat. In Africa wood represents some 86 per cent of the continent's total energy requirement, and Asia is not far behind with 64 per cent. With percentages of this order it becomes apparent that vast populations are still dependent on firewood for their day-to-day needs. Yet the amount of forested land to meet their requirements is shrinking. Huge areas of primary forest – particularly tropical rain forest – are being destroyed, and the pressure to find suitable sources of energy will increase. One estimate has suggested that if present trends continue, over 2.5 billion people will need alternative cooking fuels to replace wood by the year 2000.

At the moment rain forests cover 7 per cent of the world, some 10 million $km^2$ (nearly 4 million sq miles), which is roughly the area of the United States. However, a recent authoritative study that employed aerial photography, satellite surveys and radar, revealed that 1 per cent of the world's forest area is being devastated each year, and a further 1 per cent substantially changed. This is clearly a desperate situation, as some areas could easily lose their rain forests altogether – and soon.

The trees in the rain forests have taken decades to grow and there is a very important ecological cycle involved around them. So is there not some way in which we could grow trees much quicker on sites that are of little use to us now? Some scientists believe that they have developed an idea that may just accomplish this aim. It is a technique called 'whole tree harvesting'.

If you have ever been to a forest just after an area has been felled you may have been surprised to see how much is left behind. Large branches, small logs and trimmings are left to rot on the forest floor: but most of this can be utilized. Sweden, which has over half of its land area covered by forest, is estimated to have the equivalent of 8 million metric tons of oil in tree waste; and the same is largely true nearly everywhere.

### A wood chip boiler?
One British inventor decided to take advantage of all the waste and devised a system by which he can produce enough heat to warm water to 58°C, produce gas for cooking, and end up with an excellent fertilizer without burning a single twig. He collects all the debris, chops it up into small, wafer-thin pieces and then lets nature do the rest. For just like a garden compost heap, as the wood chips decompose they create heat. And because a pile of them acts as a good insulator, it is possible to build up enough wood chips to act as a natural self-contained boiler.

It is a rather strange-looking boiler. It consists of a huge pile of wood chips, 2 m (6½ ft) high, with a coil of piping running all the way through it. The pile, which is built up in layers, is also thoroughly soaked in cold water and this helps to prolong the biological reaction. The whole lot, it is claimed, will keep producing warm water and methane gas for 15 months; and at the end of that period the wood chips still remain useful. They can be applied as a first-class fertilizer, with a number of horticultural or agricultural uses.

Not everyone, however, wants a 2-metre-high boiler made of wood chips, even if it is landscaped and sunk into the earth to disguise its size. But trees can be used in other ways to provide energy. Instead of letting them grow for years to become mature, tall, fast-growing species can be planted and then cut down almost to a stump every four to eight years, depending on the type of tree. This ancient technique is called coppicing, and the tree is literally harvested to provide fuel. One of the most interesting experiments under way with fast-growing trees is being conducted in Ireland.

**Trees in the fast lane**

About one quarter of all the energy consumed in Ireland comes from peat, or turf as it is called there, and the country is covered in it. Irish peat comes from a time, nearly 300 million years ago, when much of the region was swampland, covered with huge tropical trees and plants. As the plants died, they fell into the boggy water where they partially decomposed but did not rot completely and over the centuries this vegetation was turned into peat. However, the peat will not last for ever. Ireland has to find a replacement fuel: supplies run out in 2020.

The country also has to decide what to do with all the exhausted peat bogs. One answer seems to be to use the bogs for plantations of fast-growing trees. For the purposes of the research the trees, which had to have a high dry matter content and grow easily on poor, wet soil, were subject to various experiments to find the most suitable species. The Irish now think they have found them and expect that if they are harvested every four years, the selected trees will continue to regrow for about 30 years (or another seven crops) before they have to be replaced.

It all sounds very promising and very simple, but for many countries there is a small question of space. Ireland is so sparsely populated that only a small fraction of the land (2.5 per cent) would have to be planted to obtain 10 per cent of the country's energy needs. However, for a country like Luxembourg to do the same, nearly half the densely populated area would have to be given up; and for the United Kingdom it would mean growing trees over virtually the whole of Scotland. In Ireland, however, since 17 per cent of the country is covered with peat bogs in any case, space is not an

Using alcohol as a car fuel is big business in Brazil.

important problem. So if all goes well Ireland should soon be meeting some of its energy needs from biomass production and, unlike the peat, this system will go on for ever.

The search for the appropriate tree to suit different climates and conditions is not confined to Ireland. In the Philippines over 12,000 hectares (30,000 acres) have been planted with a species called *Leucaena leucocophala*, or the giant ipilipil.

This tree, which matures in eight years and grows up to 20 m (66 ft) high, produces wood that burns well; and it does not mind being coppiced regularly. Its great advantage is that it can survive through protracted periods of drought, which lends itself to many Third World countries. It also does not deplete the fertility of the soil as it fixes nitrogen in the air. Its fast-growing capability can, however, give rise to problems. In Tanzania the tree has turned into a choking weed and a breeding ground of the tetse fly; in other countries it has failed to grow well on the acidic and aluminous soils commonly found in the tropics. However, given the right controls and conditions, the giant ipilipil does seem to offer good results, producing more wood than any other tree yet tested and providing good quality charcoal.

### Tanking up on alcohol

Another prospect for the future is the use of alcohol as an alternative fuel to petrol. The leading country involved is Brazil which derives alcohol from its vast plantations of sugarcane and in 1982 the country produced some 4.2 billion litres (1.1 billion US gal). The programme has had its difficulties but it is hoped that in 1984 between 6 and 7 billion litres (about 1.8 billion US gal) of alcohol will be distilled. This is the equivalent of about 75,000 barrels of oil a day, which will make a saving of some $2 billion on the country's oil bill. It is a saving that not only helps Brazil's economy, but has also transformed its car industry. In November 1983, the Brazilian motor industry built its millionth car with an alcohol-powered engine, and now more than one in ten cars are running on fuel derived from sugarcane.

Using alcohol as a petrol substitute is nothing new. Before the vast oilfields of the Middle East became the world's main source of fuel in the 1950s, cars ran on a variety of mixtures, including alcohol. Now, with the high price of oil, the old substitutes have seen something of a revival.

Essentially two types of alcohol have been employed as fuel, methanol and ethanol, and they can be used either on their own or blended with petrol. Methanol can be produced in two ways. Firstly, as wood decomposes under heat in the absence of air, the resulting liquids and tars are distilled into methanol. The second involves natural gas. And since many remote oilfields still flare off a staggering 170 billion m$^3$ (6000 billion cu ft) of natural gas every year, it would seem that this option seems by far the most commercial.

Ethanol, on the other hand, is already produced commercially from sugarcane and molasses, or from starches like cassava, corn and potatoes. It can also be made from wood, urban waste material, sewage, and even animal dung. All these materials are plentiful, so will there be a worldwide revival of alcohol-driven cars?

If a report produced by the West German government, in conjunction with Shell and Volkswagen, is to be believed then there should be. The report showed that a blend of 15 per cent methanol with petrol was an attractive fuel and after extensive tests a vehicle was even achieving better mileage on it. In Brazil, practical experience disillusioned many motorists at first when their alcohol-powered cars developed problems. However, the problems have been taken to heart by the manufacturers and Ford, for example, has changed over 100 of its 2000 components. So alcohol-powered cars should now be viable prospects.

Other oils can power vehicles as well. Soybean and a number of other vegetable oils can work as an alternative fuel for diesel engines. While research indicates that pure soybean can cause damage to engines when used for extended periods, small proportions – say 10 to 20 per cent – in the diesel mixture seem to cause no ill effects. But perhaps the most interesting possibility for using soybean oil as a diesel fuel comes when it is first converted to its methyl or ethyl esters. These esters have essentially the same volatility as diesel oil, and in all published tests it seems that they are comparable to high quality diesel oil with respect to wear, damage and overall behaviour of the engines. As a bonus, the conversion of soybean oil to these esters yields glycerine as a by-product. It is not only soybean oil that shows promise; in the Philippines experiments have been under way for some time now with coconut oil, aimed at providing substitutes for petrol, kerosene (paraffin) and diesel fuel.

Although all these materials offer some hope for the replacement of petrol as the primary fuel for transport, it will take some very strong acts of political will to achieve it. Slogans like the French 'Put a Jerusalem artichoke in your tank' may sound attractive, but many of the renewable energy plans that emerged after the first major fuel crisis have not come up to expectations.

The best that most European countries can hope for at the moment is that substitutes will be blended with petrol, and that the initial 5–10 per cent proportion of alternatives will grow. Part of the problem lies in the ecological effects of actually producing large quantities of biomass material suitable to be turned into alcohol. To contemplate turning over large areas of agricultural land to provide small sections of our energy supplies could well upset the balance of food production and lead to shortages, particularly in the Third World. To carry it out would require strong ecological management, and a good measure of social and political responsibility around the world.

### A load of old rubbish?

Perhaps one way of relieving the pressure on food-producing land may be to exploit the vast resources that we throw away daily – our waste materials. Over 30 million metric tons of rubbish will be dumped in Britain in 1984 with a calorific value equivalent to 12 million metric tons of coal. And only 20 per cent of this is not highly combustible. But even that 20 per cent could still be useful since most of it is glass or metal, both of which are eminently suitable for recycling.

The United Kingdom Department of Energy's Technology Support Unit (ETSU), based at Harwell has done detailed research into the savings to be gained from waste, although its figures were more conservative. The ETSU study showed that about 5 million metric tons of coal could be saved by using waste as fuel. In terms of hard cash, that amounts to about £300 million a year at 1983 prices in reduced fuel costs for industry, commerce and the public sector. The long-term potential savings are even greater, as

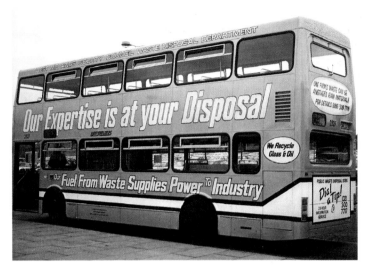

ABOVE LEFT AND RIGHT In England the West Midlands County Council has invested in waste fuels. The Coventry Waste Reduction Unit, for example, exports heat energy to local industry 24 hours a day, every day of the year, and provides over £700,000 income to the County Council. The council's fuel is known as 'Econofuel' and it spreads the word using the local buses

there is also the added benefit that there is less waste to dispose of, easing the pressure on our environment. Besides this, it produces considerably less sulphur dioxide than conventional fuels.

The idea of obtaining energy from waste materials, like most of the renewable technologies discussed so far, is not revolutionary. There was, for example, an incineration plant in Glasgow which produced electricity from refuse as early as 1912. It continued to operate until the early 1950s when the advent of cheap liquid fuels led to a decline in interest. The use of waste was, however, kept up in the furniture industry, where the use of dry wood waste has always been commonplace.

Waste materials have properties that set them apart from the more conventional fuels such as coal, oil or gas. Generally they have a lower calorific value, a higher volatility and a lower fixed carbon content than coal, and often come in irregular and awkward sizes. They are usually much less consistent than conventional fuels and need to be burned on specialized combustion equipment. Nevertheless the technology involved is not difficult, and there is a wide range of options. Domestic refuse, for example, can be burned directly in a large specially designed incinerator; or it can be extensively pretreated to produce a refuse-derived fuel that can be burned in a smaller, simpler and cheaper furnace.

A number of schemes using refuse are already operating. Liverpool is, at the time of writing, planning to build a new plant at Huyton that will process 300 metric tons a day. The refuse will be fed through a fine vibrating screen which separates out the smallest, organic particles, These can be used separately as compost for treating waste land that is being reclaimed. The heaviest part of the refuse, mainly metals, is sorted next and removed for separate reclamation, leaving the lighter part, including paper and textiles, to be pressed into pellets. These are dried, cooled and hardened before being sold locally as fuel. Elsewhere in England, at Westbury in Wiltshire, refuse from the local authority is also being used, this time to power a cement works.

Perhaps the real potential of using waste as fuel can be seen from the results obtained at a large plant at Edmonton in London, and at other installations around Europe. The Edmonton plant has been working since

1974 and operates continuously, burning about 400,000 metric tons of refuse a year and using it to create electricity. Although it burns less than 2 per cent of Britain's refuse, it has already saved about 1 million metric tons of coal in nine years.

If this plant could operate on a combined heat-and-power basis, it would be able to offer even greater savings. Nor is Edmonton the only plant with good operating experiences. Two incinerators at Bernard Road, Sheffield, are coupled up to boilers to produce steam, and this provides heating for a block of flats. In West Germany at the Flingen refuse power plant at Düsseldorf six units supply steam to a local electricity power station, and for district heating. Altogether there are some 200 waste-burning plants all over western Europe and the idea has also spread to the United States. In Saugus, Massachusetts, a huge waste-to-energy plant was constructed and now handles an average of 1200 metric tons a day. Following this success another such installation is planned in New York. This will be able to cope with 2250 metric tons a day and will produce 60 megawatts of electricity for sale to the local generation utility.

The plants do not even have to be fixtures. One boiler, appropriately christened the 'Energy Capsule', comes fully containerized and self-contained, complete with fuel silo, water tanks, chimney and connecting pipework, and it is totally portable. Within 48 hours it can be connected up to the existing heat distribution system and have steam or hot water on line.

### Recycling

Could we save more by recycling materials? Recycling became rather a fashionable thing to do some years ago, but now seems to have lost ground a little. Most people seem unaware of which materials can be re-used, and little official encouragement is given to separating domestic refuse into its component categories for recycling.

Recently, however, the importance of recycling materials like aluminium, iron and wood has again emerged. These three items are now among the world's most vital materials, and it has been calculated that less energy would be used if we could obtain our stocks of them from waste, instead of exploiting new sources. Since throwing away one aluminium drink can wastes as much energy as emptying a can half-filled with petrol, the point can readily be taken. In fact, only about one quarter of the world's paper, aluminium and steel is at present recovered for re-use.

It is to be expected, perhaps, that countries rich in materials and resources, or countries in which cheap energy is readily available, are reluctant to develop recycling techniques. Those countries that have responded have only done so either because there has been a shortage of raw materials, or their energy and capital costs for processing new materials have been high, or because the environmental damage was beginning to increase. Japan and the Netherlands, for instance, now lead the way in waste-paper recovery. Japan uses half of what it once consumed, and the Netherlands meets about 40 per cent of its paper requirement by recycling. One Japanese city has even gone so far as to recycle about 90 per cent of its garbage.

The United States, with its emphasis on consumerism, has a rather different record to that Japanese municipality, particularly regarding aluminium. America actually throws away more recyclable aluminium, in the form of drink cans, than the whole of Africa produces. When you realize that it needs 20 times as much electricity to smelt aluminium from its ore, bauxite, than to produce it from recycled material, the scale of the savings can be appreciated.

**Biogas**
The idea of refuse-derived fuel can be taken one stage further. There is strong interest now in exploiting deposits of methane gas, which is generated by the natural decomposition of organic waste buried in huge rubbish dumps and landfill sites. The gas, which occurs naturally and is identical to that extracted from expensive wells beneath the North Sea, often seeps up to the surface causing 'bad egg' smells and small explosions.

If you visit Stewartby in Bedfordshire you will see methane gas being harnessed. Up to ten gas 'wells', placed at 50 m (55 yd) intervals across the surface of a landfill site, have been established and gas is continuously extracted. It is then cooled to remove any water present, before being compressed and piped to a nearby brick tank where it is burned to replace coal. Another such project is working well at the Greater London Council's site at Aveley, Essex. The results so far have been promising enough to make the waste-disposal industry contemplate redesigning its landfill sites as huge 'bioreactors'. The Department of Energy's Technology Support Unit suggests that the 20 largest suitable landfill sites in Britain are capable of producing gas with a heat equivalent to over 300,000 metric tons of coal a year, so it is clear that this 'biogas' is likely to be more widely utilized. In the United States, where some landfill operations receive more than 5000 metric tons of refuse a day, an estimated 1 per cent of the country's total energy needs could be met from this resource if the sites were all tapped efficiently.

If the gas is so valuable is there not some way of speeding nature up? The potential from refuse gas has prompted some scientists to do just that and develop cultures of bacteria and other micro-organisms to seed into the dumps. By implanting a whole cocktail of 'bugs' in the sites it is hoped that they will rot down faster and produce their gas quicker and more efficiently. The bacteria are carefully selected from cultures taken from refuse dumps and other sources, and combined into communities of perhaps a dozen species apiece, which work together to get the best performance out of the waste material.

The gas is not just being produced at landfill sites. It can be extracted from farm waste, domestic sewage or industrial effluents through anaerobic digestion. Anaerobic digestion is the name given to the complex, naturally occurring bacterial process by which organic matter is decomposed in the absence of oxygen. In nature the process occurs in stagnant ponds, marshes and in the mud of polluted rivers. By building a digester to turn farm slurry into methane, one EEC-backed experiment in Belgium has not only depolluted and deodorized the waste, but also supplied enough energy to meet the electricity and heating needs of the particular farm. Any surplus gas has been employed to heat greenhouses and the final residues have been used as a stabilized fertilizer, easily assimilated by the plants.

The advantages that come from anaerobic digestion of farm waste could be exploited all over the world. Animal wastes such as cow dung are common to all farming communities and small biogas plants could be installed on a village basis, improving the lot of many Third World areas and reducing the dependence on firewood. India, where the end product is known as *gobar* gas, is one example where cattle are widespread and

numerous, and the potential for such plants is promising. There have been a number of problems, however, in getting these digesters to work and few benefits have materialized yet to solve the energy problems of India's poor.

The Chinese, by contrast, seem to have made a success of the process and reported to a recent United Nation's Energy Conference that they had over seven million such digesters in operation. Concentrated mostly in the central province of Sichuan (Szechwan), the plants apparently supply methane to 30 million people. There are also some 150 methane-fired power stations in operation, a considerable feat.

One of the problems that the Chinese will have faced in making their power stations work is one that has dogged early researchers in this field. For the biggest headache in burning untreated biogas is that trace acidic gases are produced and these have a severe corrosive effect on components like copper. This is particularly evident where gas engines are used inter-mittently, allowing a condensation of gases to occur. A number of trials are being conducted in Britain at the moment in order to study the problems of corrosion. No results will be available for some time yet, but initial tests appear encouraging.

Cellulose is a resource that is almost as common as waste and it, too, can be used to produce energy. Over $10^{11}$ metric tons of it are produced every year. Through a variety of reactions a number of chemicals like ethanol, acetone, citric or butyric acids, etc. can be made more conveniently and cheaply from biomass-derived intermediaries than from petroleum. According to Professor David Hall of King's College, London University, 'microbial technologies are now competitive in the production of high tonnage chemicals when applied in conjunction with the disposal of wastes'.

Is it therefore possible to build a self-sufficient biomass community, a photosynthesis energy factory? From the evidence available it does seem that if a factory and agricultural production system were combined, non-polluting and totally domestic fuels derived from solar energy and the various residues could provide energy on a renewable basis.

ABOVE LEFT The Coventry Waste Reduction Unit
ABOVE RIGHT Vast amounts of 'rubbish' is thrown away as waste; but it could be used as fuel

RIGHT Landfill gas being
flared at Stewartby,
Bedfordshire
BELOW The water hyacinth –
how to turn a rampant weed
into a useful energy source

Aquatic plants could also be used to develop the system further and a great deal of work has been done on seaweed kelp with the idea of growing it to be converted into methane. The plan was to grow it on millions of acres of floating platforms at sea, but the project has yet to prove feasible.

The water hyacinth is another possible choice. Normally a pest because it doubles its size in eight days, clogging up waterways and lakes, it can produce 1 kg of biomass per 1 $m^2$ of water per day, and can survive even in relatively polluted water. Cassava and cattail, two other weeds, are also good contenders. Using these plants on marginal land could make the self-sufficient biomass community a more widespread phenomenon, and one that would not necessarily involve using up good agricultural land.

## Reinventing photosynthesis

The ultimate method of creating renewable biomass energy would be artificially to mimic the process of photosynthesis. It would amount to solving with chemistry a problem that nature solved aeons ago – the use of sunlight to split water into oxygen and hydrogen. The reward would be a cheap, clean and inexhaustible fuel, hydrogen, and no plants or forest would have to be destroyed.

In 1874 Jules Verne wrote of his belief that 'water will one day be employed as fuel, that hydrogen and oxygen which constitute it used singly or together will furnish an inexhaustible source of heat and light'. The leading scientists in the field today would probably echo his thoughts. But although some progress has been made in the last ten years, scientists are still a long way from being able to isolate hydrogen from water on a commercial scale. It is not difficult to turn water into hydrogen and oxygen; it can be done by using two electrodes and passing an electric current between them. But it is an extremely inefficient process, using up valuable energy in the electricity consumed, and is not an economic proposition. Researchers instead would like to emulate nature's successful use of solar energy to produce hydrogen biologically.

A number of experiments are being conducted, looking at the problem in different ways. One complex but promising method is to create hydrogen via a chemical reaction, using metal catalysts like platinum and phosphorus. Another is to exploit nature's ability to use sunlight to separate positive and negative charges. The basic photoelectrochemical system is the key to a photosynthesis, and depends on the fact that each organism has a membrane that contains chlorophyll. The idea is that by creating artificial membranes containing chlorophyll, it would be possible to stimulate the release of both oxygen and hydrogen. A more natural method involving plants is to use certain algae that produce hydrogen under specific conditions and all of which contain the enzyme hydrogenase. What is now being examined is whether plants can be stimulated to produce hydrogen by the addition of hydrogenase.

All of these techniques are still in their infancy, but the development of renewable energy sources from biomass is not. This is an energy resource capable of being harnessed to improve the quality of life of many people in the Third World. This resource, whether it is in the form of coppicing trees, or utilizing the huge quantities of refuse we throw away every day, or using plant-derived oils to reduce our dependence on finite fossil fuels, can be developed. Used responsibly and with a careful regard for the environment, biomass energy must be one of our hopes for the future.

CHAPTER SIX

# Blowing in the Wind

To suggest that tomorrow's world could be powered by devices that can prove their pedigree back over the last 2500 years might sound rather foolish. However, wind power and the use of windmills to harness it is now being seriously considered as one of the most promising of the renewable energy technologies. Today's designs do look rather different, bearing only a distant family resemblance to the corn-grinding and water-pumping machines of the past. The new wind machines owe more to the lessons of aerospace engineering and of our increased understanding of materials than to their ancient predecessors.

Man has been trying to exploit the power of the wind for centuries. The earliest mention of windmills in actual use comes from India about 2500 years ago (and, of course, sails were used on ships a lot earlier than that). In Persia, Iraq, Egypt and China many early designs were produced. Taming the wind provided, with waterwheels, the main alternatives to muscle power for high quality mechanical energy.

By the early nineteenth century windmills played an important role in Europe's economic life. Before the Industrial Revolution for example, there were something like 10,000 machines operating in Britain, and similar figures were common to other European countries. One 25 m (82 ft) rotor made from wood and canvas could do the work of more than 200 men or women. But in terms of the amount of energy windmills produced out of the wind, they were still very inefficient. With the dawn of the Industrial Revolution, any early efforts to improve their efficiency and performance were overtaken by the pace of technological change along other lines.

As steam engines burning fossil fuels appeared and became economically more attractive, the advantages they enjoyed in being able to provide mechanical power on demand heralded the demise of the windmill in Europe.

However, not everyone turned their backs on wind power and research did continue. Windmills began to be designed specifically to produce electricity. By the end of the nineteenth century the first wind power stations were coming off the drawing board. The first units came into service in the 1890s in Denmark, and by 1908 several hundred small stations were operating. They produced very small amounts of power, between 5 and 25 kilowatts, which was only sufficient to power a few homes, but they ensured that wind turbine technology survived.

The technology was extended still further by the appearance of larger machines like the one at Yalta in the Soviet Union. One of the first on the scene, it was rated at 100 kilowatts and was essentially a very basic device. Its blades were covered with roofing metal and its main gears were made of wood – but it worked. Development work on other designs continued after the end of World War Two. In France, for example, there was an 800 kilowatt plant connected to the grid at Nogent-le-Roi, near Paris, and others were built at St-Remy-des-Landes in the southwest and on the island of Ouessant, off the Brittany coast. In Denmark, the Netherlands, Germany and Britain more turbines were constructed. Some had two blades, others three, but their development proved short-lived. With the advent of cheap oil in the 1960s, they were abandoned.

However, the influence of these postwar designs has remained, affecting the technical lines of the wind energy converters of the late 1970s. For wind power has been 'rediscovered' following the energy crisis of 1973, and development in this direction looks this time to have a somewhat longer and more durable future.

### Just one per cent?

Wind energy is an indirect form of solar energy. It results from the fact that the earth's equatorial regions receive more solar radiation than the polar regions, which causes large-scale atmospheric convection currents. About one per cent of the world's incidental solar radiation energy is converted in the atmosphere into wind energy; this may not sound very much, but it is in fact a vast and widely distributed resource. To quantify just how much wind could be utilized for electricity generation, an extensive study was carried out for the EEC. It concluded that even after allowing for the many constraints concerning where these large windmills can be placed, there were sites for about 400,000 megawatt-scale turbines in Europe; enough, in principle, to provide about three times Europe's present electricity consumption.

Unfortunately there are some potentially serious drawbacks. The wind is by its very nature intermittent and so, too, is any power that it can generate. So whatever power is produced cannot be guaranteed to be available in, say, a few hours' time: not the best way to fuel Europe's industries, you might think. But with national grids supplied by existing conventional sources, wind generation could provide at least some of their electricity and save on expensive and renewable fossil fuels.

Large-scale electricity generation using wind converters involves high-technology design and components. It is at the moment in its developmental stage as scientists and engineers wrestle with some of the large problems involved. However, small-scale wind turbines are already proving themselves to be both reliable and economical. Small windmills for pumping

BELOW LEFT One of Britain's new wind turbines in the Orkneys
BELOW RIGHT Windmills have been used for centuries to provide mechanical energy. Today's designs to produce electricity look rather different

water, for example, are still widely used – particularly in Australia and the
United States – and it is estimated that more than a million are currently in
service. Some of the largest machines are in Australia, with rotors reaching
up to 10 m (33 ft) in diameter. Using light winds, these are capable of lifting
water some 200 m (655 ft) or more. Generally used in remote areas, these
devices are making a particular impact in developing countries where
water is often scarce.

## A fair wind?

Small wind generators can also be used to produce electricity, and again it
is the smaller, more isolated communities, which normally rely on expensive diesel-fuelled plant, that can benefit the most. One British example is
Fair Isle. The inhabitants of this remote island between Orkney and Shet-
land have had their life styles considerably improved by the installation of
a 50 kilowatt wind generator. By developing an ingenious load-related
control system for distributing the power, they have managed to slash their
electricity costs from over 13 pence per kilowatt hour (when they were
using diesel) to just 3 pence per kilowatt hour. This price also includes all
the costs of operation, insurance and maintenance on the turbine. The
diesel unit is still used to maintain essential services and to operate when
the wind generator is becalmed, but three-quarters of this northerly island's
heating, lighting, hot water and electricity comes from the wind.

Fair Isle is not just a one-off; international studies have identified about
200 islands throughout the world where diesel generators supply all the
electric power with a minimum load demand of 100 kilowatts and upward,
and an average annual wind speed greater than 5 m per second. There are
also many hundreds of remote communities in countries with large land
areas. In Canada, for instance, about 100 out of 268 remote communities
should have a good potential for wind power. For as long as the area has
enough wind, and 5 m per second is about the minimum required, and
provided it has the back-up of a bank of batteries for storing surplus power
to be used when the turbine is not turning, or of standby generators
powered by fossil fuels, large savings can be made.

Rural and isolated areas are not the only ones who could benefit if
medium- and large-scale turbines become as economically successful as
the small versions. A firm contribution, regardless of the vagaries of the

weather, could be made to a national grid. In Britain the present electricity grid could accept an input of 20 per cent from wind power. If slow-response steam turbines were eventually replaced by low-cost gas turbines, able to respond very quickly if the wind dropped and the wind stations began to reduce their output, it has been suggested that even greater savings could result.

**California's wind valley**
California has already taken wind power to heart and integrated some wind turbines into its electricity grid: 300 megawatts to be exact, and all coming from small- and medium-sized machines. Such is the level of enthusiasm that two big Californian electricity companies have announced plans to double the capacity at least by 1990. And it is not just the large companies who are catching the 'wind bug'. The situation has now developed to such a stage that agricultural land has been turned into 'wind-farming' areas, with 'prime wind land' being offered to farmers where cattle once grazed. The best-known and most successful of these 'wind farms' is at Altamont Pass, where harvest times run from April to September. (In California, the windy season begins in the summer, neatly coinciding with the peak demand for electricity as people switch on their air-conditioning systems.)

As the temperature starts to rise in the central Californian valleys, cool heavy air from the coast is sucked along the natural funnel of the pass and it blows consistently enough to provide (with the help of some good tax credits) commercially viable wind power. In the past two years more than 2000 microprocessor-controlled wind turbines have been erected along the pass by private companies, and hundreds more are on the way. Altamont provided one of the largest wind operators in 1983 with nearly 23 million kilowatt hours of electricity to sell to the local utility. In three of the big passes in California there are now some 10,000 machines operating.

**A technical hitch – or two**
However, the wind turbines at Altamont, like many others, have not been without their fair share of technical problems. The scale of the difficulties has been related to the size of the machine, but some of them are common to all. Part of the problem lies in trying to make efficient devices capable of generating power with variable wind speeds. One leading researcher, Dr Louis Divone, head of the American wind energy programme, has likened the loads a windmill designer expects his machine to cope with, and survive, to asking an old bomber to do continuous barrel rolls in turbulent air 50 m (165 ft) off the ground for 30 years. One of the worst anxieties of the aerospace engineer is fatigue, the weakening of material by endless fluctuations in loading; for the windmill the fatigue problem is much more serious.

In the United States, and particularly at Altamont, the machines vary greatly in design and reliability. A great deal of work has gone into just learning, as it were, to live with the machines and using the experience gained with the prototypes to design better successors. This is particularly true for larger machines. One Darrieus-style turbine (of which more later), which was designed to withstand winds up to 160 km/h hour (99 mph), was caught by a 65 km/h (40 mph) gust while starting up during tests in southern California and collapsed. The fate of a horizontal axis turbine on the Île d'Ouessant in France was no better. The machine, rated at 100 kilowatts, decided to dispose of one of its blades, depositing it 180 m (200 yd) away.

### Converting wind into energy

Wind has energy because of its motion and windmills gather it by slowing down the mass of moving air and converting it into mechanical energy. As a result of that process there are three obvious and important factors: the speed at which the wind is blowing; how much wind the rotor can intercept; and the overall conversion efficiency of mechanical parts such as the rotor, transmission system and generator. But however efficient a windmill is, it can only convert on average about 30 per cent of available wind energy into mechanical energy.

One prerequisite for any good wind turbine site is that it should be windy. This is not meant as some kind of joke: if you double the wind speed, the power produced is multiplied by eight times. (The power that is available in the wind increases as the cube of the wind speed.) As the wind speed varies, so too will the output of the turbine. Most wind generators, therefore, have been designed so that they cut in at a certain low wind speed and then steadily build up as the wind increases to their rated output. Once they have reached this point they will not increase the amount of electricity they produce any further. The difficulties and the costs involved in designing a useful application capable of absorbing what may well be drastic variations in the power output are, at the moment, just too big.

The problem becomes even more acute when the designer is faced with very high wind speeds. Most machines now have an automatic self-preservation mechanism, so that when the winds reach storm force, the machine shuts itself down and prevents any risk of overrunning. For most turbines, the operating range lies between 5 and 27 m/sec, or between Force 3 and 10 on the Beaufort Scale.

**The Beaufort Scale**

| Beaufort number | m/sec | mph | Description |
|---|---|---|---|
| Force  0 | 0–0.5 | under 1 | calm |
| 1 | 0.5–1.5 | 1–3 | light air |
| 2 | 2.0–3.0 | 4–7 | light breeze |
| 3 | 3.5–5.0 | 8–12 | gentle breeze |
| 4 | 5.5–8.0 | 13–18 | moderate breeze |
| 5 | 8.5–10.5 | 19–24 | fresh breeze |
| 6 | 11.0–14.0 | 25–31 | strong breeze |
| 7 | 14.5–17.0 | 32–38 | near (US moderate) gale |
| 8 | 17.5–20.0 | 39–46 | (US fresh) gale |
| 9 | 21.0–24.0 | 47–54 | strong gale |
| 10 | 24.5–28.0 | 55–63 | storm (US whole gale) |
| 11 | 28.5–32.5 | 64–72 | violent storm (US storm) |
| 12–17 | 33.0 and over | 73–136 | hurricane |

But how are the turbines able to produce the same power if the wind speed increases? In most designs the blades are 'feathered' automatically, their pitch being changed in response to the wind speed. In the very large machines this adjustment is computer controlled: sensors are alert to alterations in either the wind speed or the electrical output and change the pitch of the blades accordingly. If the machine needs to be stopped then powerful brakes are applied. These can bring the rotor down from a speed of around 27 m per second to a standstill in about two seconds.

The size of the rotor blade is a critical factor in any wind turbine. For if the diameter of the rotor is doubled, then the swept area and the power output is quadrupled. This formula applies to the most popular and established design of wind turbine, the horizontal-axis type, which can have any number of blades. Often referred to as HAWT for short, these machines are virtually direct descendants of the windmills of the past. In most versions the rotor is mounted on a horizontal shaft and connected to a generator; and like most turbines, the taller the tower it is mounted on, the better the performance, for wind speeds increase with height.

### Horizontal axis wind turbine

The development of horizontal-axis machines for generating electricity in recent years has been pioneered by the United States. However, the Americans have certainly not had a monopoly of the technological advance in this field. The Danish government launched a big programme in 1977 and two 630 kilowatt turbines have been built at Nibe in Jutland. The emphasis there has been on three-bladed machines and it has been estimated that 10 per cent of Denmark's electricity could be generated from the wind by the late 1990s. Other European countries like Sweden and West Germany have also constructed large turbines. But perhaps some of the most exciting and optimistic developments for large-scale wind turbines are taking place in Britain.

### The Orkney experience

If you were looking for a site to place a wind turbine somewhere around the British Isles, then Burgar Hill in Orkney would be near the top of any list. Power is expensive on the islands, where electricity comes mainly from diesel-powered generators. The site, too, is fairly remote and best of all, it is windy. It is in fact one of the windiest sites in the country. The mean annual wind speed which a small machine might face is estimated at 10.5 m per second; for a larger machine, such as a 3 megawatt design, the figure is even higher and reaches 12.6 m per second. It is not too surprising, therefore, to find that Burgar Hill has become host to two large British turbines and will soon have a third.

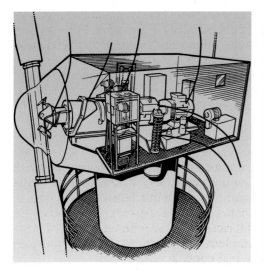

FAR LEFT A closer look at the diagram shows just how large it is
LEFT The Howden turbine, Orkney

The two turbines, which are made by different companies, may look similar but are designed to achieve different aims. The smaller of the two has a 20 m (66 ft) diameter rotor and produces 250 kilowatts in a near gale force wind of 17 m per second, and it rotates at 88 revolutions per minute. It is a prototype machine, serving as a research tool for a larger 3 megawatt machine to be built on the same site in 1985 and 1986, and also functions as a generator in its own right. The two-bladed rotor produces enough electricity to supply 150 homes, and is mounted on the top of a part-concrete, part-steel tower. It will have an annual output of about 700,000 kilowatt hours and its rotor will turn about 30 million times a year.

As with all horizontal axis turbines, the two-bladed rotor, which mimics the action of a giant aircraft propeller, needs to be kept turned into the wind. For unless the wind tends to blow consistently from one direction – as does happen in a few rare locations – the turbine's rotor must be mounted on some kind of turntable. A series of sensors are normally fitted, which signal an elecric motor to keep the rotors constantly facing into the wind.

This 250 kilowatt turbine has some rather remarkable additional features. A sophisticated microcomputer control system enables it to operate in two entirely different modes. In the first mode the rotor is run at a constant speed directly connected to the grid. In this case the control system keeps the rotor speed constant at 88 revolutions per minute as the wind speed changes by pitching the blade tips, i.e. keeping the blades at the same speed while the wind speed increases. The arrangement is rather like the flaps fitted to aircraft wings, with the rotor blade tips generating levels of reverse thrust at high wind speeds.

In the second mode the rotor is allowed to run at variable speeds in the range of 44–88 revolutions per minute so as to extract maximum power from the lower velocity winds. Again, the rotors are prevented from straying beyond these limits by using the blade tips as correcting forces. In this mode the generator produces a variable-frequency output, which needs to be rectified and altered to give the fixed frequency that is needed to synchronize with the grid.

The small 250 kilowatt machine has already proved that it can be reliable and valuable both as a generator of power and as a research tool. It has been constantly monitored by a complex system that handles some 200 signals from a range of transducers, including crack-detection gauges, position sensors and strain gauges. It has endured many storms, lightning strikes and occasional severe icing conditons, all of which have created a range of opportunities to test the machine's design and control system. The fact that it has survived them and is working well has boosted confidence for the development of the 3 megawatt turbine, which is a scaled-up version of the 250 kilowatt design. The two machines together will supply 15 per cent of the island's electrical needs.

Another small part of Orkney's electricity is generated by the other major British offering. This turbine has been specifically designed to have an immediate commercial role and not as a research tool for larger, more ambitious projects. The machine, which generates 300 kilowatts, is available in two versions: the first is suitable for wind farms; the second is designed specifically for use in isolated communities or on islands like the Orkneys. One of its features is that it can be erected without the use of large cranes. This may sound rather mundane but the logistics of getting large cranes to remote islands is a very real consideration.

Orkney is not the only site, however, for British wind power developments. A commercial-scale machine is in place in Devon, and an important experiment has also been under way at Carmarthen (Caerfyrddin) Bay in Wales, where the Central Electricity Generating Board (CEGB) has installed an American-designed 200 kilowatt turbine. Its annual electrical output was expected to be about 380,000 kilowatt hours a year, but the machine has been beset by problems. The CEGB has also said that it intends to buy a large machine with a rotor diameter of up to 90 m (295 ft) and a rating of up to 4 megawatts, which it hopes to build near to one of its power stations at Richborough in Kent. If all goes well, this is planned to be the first of up to ten machines to be built, forming Britain's first wind farm.

### Vertical-axis turbines

As with most scientific inventions in their early years of development, there is no consensus yet as to what constitutes the best design for wind turbines. Wind turbines of the horizontal-axis type may differ in a number of ways: they may have two or three blades, running upwind or downwind, and all are positioned on towers that can themselves be made up of a variety of materials. There are also designs of totally different types. The most advanced of these are 'windmills' whose blades operate vertically.

This type is known as the vertical-axis design, and the diversity within this group is just as pronounced. One machine is more reminiscent of an inverted eggbeater, and another of an old-fashioned television aerial, than the traditional notion of a windmill.

The first known modern vertical machine was proposed by a Frenchman, Georges Darrieus, in 1931, but it was not developed because of a general lack of interest in wind power at the time. The design did not re-emerge from academic obscurity until the 1960s, when the National Aerodynamic Laboratory in Ottawa, Canada, built and tested a series of machines. But why bother with the design at all?

By using the vertical axis, the turbines do have some important advantages. They do not need a control mechanism to steer them into the wind: they can by virtue of their design react to whatever direction it is blowing in; they do not need as much structural support as the horizontal versions because heavy equipment like gearboxes and generating machinery can be placed near to ground level; finally, the rotor is not subjected to the same continuous cyclic gravity loads that can cause fatigue in the horizontal design, as their blades do not turn over end over end.

Vertical-axis turbines do have their disadvantages in that most of the designs are not self-starting and have to be turned before they will generate enough lift to produce power. They are prone to more severe problems of vibration and stability than the horizontal designs. However, in terms of technological development, they are much younger machines. Horizontal-axis turbines have been under scrutiny for far longer, and with a much greater intensity of effort, than the vertical types, so their present faults may well be soon resolved.

To say a wind turbine looks similar to an inverted eggbeater does not sound very complimentary, but it does aptly describe the appearance of a Darrieus-style machine. The basic configuration uses two or three bow-shaped blades which are connected to the top and bottom of a central vertical shaft. As the wind flows through the rotor, it turns the aerofoil-shaped blades, so driving the shaft to produce power.

The Darreius turbine

There are a number of small and medium-sized vertical-axis designs now available on a commercial scale, but most of them are still being built for experimental purposes. One of the largest is a 230 kilowatt Canadian machine which has been built on the Magdalen Islands in the Gulf of St Lawrence, but it will soon be superseded by much larger machines in various countries, including one rated at 4 megawatts which is due to be operational in 1985.

### The Musgrove turbine

While development of the Darrieus turbine is continuing, another vertical axis project, under way in England, has begun to cause a great deal of interest. It was designed at the University of Reading, Berkshire, by Dr Peter Musgrove, who was inspired by work done in Canada. This H-shaped machine was proposed because vertical-axis wind turbines are highly efficient at wind speeds below 32 km/h (20 mph), but once that figure is passed they run into problems. Firstly, it is more difficult to control the power output in high wind speeds than in a horizontal-axis machine. It would have to be varied cyclically every revolution, which would introduce considerable mechanical complexity. Secondly, high wind speeds could generate enough centrifugal force to push the tips of the blades outwards, and could become strong enough to break up the windmill. In Canada they tried to use curved blades attached to the vertical centre pole at both ends to neutralize the centrifugal effect. This did not eliminate it entirely. In high winds the centrifugal force would be capable of tearing the whole blade away from its mountings. To avoid a disaster like that the Darrieus turbine would have to shut down in such conditions.

At Reading a much simpler technique has been used to produce a machine that can go on generating through the strongest gales. Instead of having curved blades like the Darrieus, the Musgrove version has straight, untapered and untwisted blades. These form the two vertical outer sections of the H-shape and are connected to a horizontal bar that joins them together. They remain in that position when the rotor is stationary, or when it is turning in low wind speeds. However, when the wind speed begins to pick up, the design reveals its unique features. It can vary the geometry of its blades. So as the centrifugal force increases, the blades no longer remain in the vertical position, but pivot outwards. How much the blades actually move towards a < shape is controlled by wires which attach the blades to a large spring inside the main central support. As the wind speed increases the tips of the blades, under the influence of centrifugal force, are able to lean out, forcing the spring to extend. Although the overall efficiency of the turbine drops as the blades fold, it does mean that the machine does not have to shut down every time the wind becomes strong. The Musgrove version also has the added benefit that its straight blades are easier to manufacture than its curved Darrieus counterparts.

The Musgrove machine has now left the drawing board and now that all initial trials on small prototypes are over, a full-scale 'model' is being constructed close to the CEGB's existing horizontal-axis machine at Carmarthen Bay in Wales. It is a 'model' in that it is acting as a testbed before a much larger machine, capable of producing 4.4 megawatts and with a rotor diameter of 100 m (328 ft), can be built. The smaller version will be able to generate 160 kilowatts of electricity from a 25 m (82 ft) diameter rotor and will be commissioned in 1985.

Not surprisingly the Musgrove design is not the only one to emerge in the vertical-axis field. Another British contribution has come from the Open University and Dr Derek Taylor. Dr Taylor's turbine still has the advantage, shared with all vertical-axis machines, that it does not matter which direction the wind decides to blow in: but it also has some major differences from other devices in the category. The machine will be self-starting and will only require a low tower, about 3 to 4 m (10 to 13 ft) high, for its support, which will save on cost and be less intrusive from an environmental point of view. The blades, together with the tower, look as if they were modelled on a stationary helicopter. They are not horizontal, however, but placed instead at an angle of 45 degrees to the shaft.

At the moment the Taylor version has only been tested on a small scale. These tests have been taking place with a 1 m diameter machine in the wind tunnel at Queen Mary College, London University. A theoretical computer model of the turbine's aerodynamic performance has also been developed and wind tunnel tests show that this model can be used to design larger-scale versions. So if all goes well the turbine should be built in some prototype form by the end of the decade.

### Oblique-axis turbine

However, Britain certainly does not have the monopoly on innovative projects when it comes to generating power from the wind. One rather extraordinary development has been the oblique-axis turbine, which has been worked on by Dr Gunther Wagner in West Germany. The design, which has one main blade, is totally different to the vertical- or horizontal-axis machines. The blade is placed at an angle of about 50 degrees to the shaft of the rotor, which is itself aligned between 45 and 55 degrees above the horizontal. This means that as it is facing the wind the blade sweeps out a path from the horizontal progressively up to the vertical and then back again to the horizontal. (See illustration p. 96)

One possible disadvantage might seem to be that the turbine only generates power on its upward sweep as it faces the wind. The lower arc does not produce any at all, creating significant peaks in the amount it creates on each revolution. So because the entire output has to be obtained from only a part of each revolution from a single blade, that blade has to have a very large span. This is the case on a 250 kilowatt prototype that has been tested. It has one single effective blade, which does all the work and is balanced by a very much shorter one acting as a counterweight.

Three small Wagner-pattern turbines have been built on land, and an array of ten will actually soon be in place on an island in the Elbe estuary. The biggest application so far, however, is the conversion of a small coastal vessel to carry a wind turbine. The blades sweep up and over the ship and then come down parallel to the foredeck. The result is visually very very striking.

The ship now serves as a prototype for an ambitious project. The idea is to use ships as vehicles on which to place wind turbines at sea, instead of having to mount them on the equivalent of North Sea oil platforms. Undaunted by the complexity of the design, there are plans to build very large Wagner rotors of about 7 megawatts on ships and to place them out to sea. These would be connected by cable to the island of Sylt off the northwest coast of Germany. The electric current produced would then be fed back to the mainland.

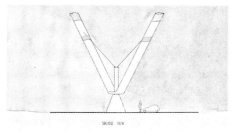

RIGHT The Taylor vertical-axis wind turbine: an artist's impression
FAR RIGHT A model of the turbine under test in a wind tunnel

### An environmental hazard?

The idea of having clusters of wind turbines out at sea is not confined to the work of Dr Wagner. It has been developing for some years now and looks like becoming a realistic prospect some time in the early 1990s. But why place wind turbines offshore where they are more difficult to operate and maintain, and are more expensive to install? Part of the reason lies in the environmental problems that wind turbines can cause when they are sited on land. The revolving blades generate a certain amount of noise, although it does not travel far, and they can also cause interference with television and radio signals over a limited area. Again, the problem is not insurmountable since cables could be laid to carry the signal, but this would all add to the net cost of the electricity that would be produced.

Nature could also add to this list of problems by the severity of the weather conditions. In very cold weather, icing has been known in the past to cause severe difficulties and there are cases where the rotors have over-speeded and broken up. The risk of a turbine blade flying through the air is extremely remote, but the consequences could be very distressing indeed. However, perhaps the greatest problem is caused not so much by nature but by man's love of the natural beauty of the countryside.

Huge windmills of this modern kind are visually intrusive and since most of the best sites for wind power are in areas of outstanding natural beauty, the environmental costs could well be thought to be excessive. At least 1500 hilltop sites around the British Isles have been identified as being suitable for wind turbines. These spots, of which only 320 are in England and Wales, have a high average wind speed throughout the year (around 9 m per second). If you could install a large machine, like the 60 m (197 ft) one proposed for Orkney, on all of the sites, then wind power would certainly make a very positive contribution to Britain's energy requirements. Theoretically, wind power could provide the equivalent of some 7 per cent of the present British electricity supplies. Dr Peter Musgrove has gone further, estimating that the potential contribution from land-based machines could be at least 10 per cent. But that would require something like 3600 turbines, each of 2.5 megawatts, in groups of 25; each group would be within an area of 4 km by 4 km (2½ miles by 2½ miles). This amounts to some 2300 km² (890 sq miles), or about 1 per cent of the rural land in England and Wales. Would we be prepared to give up this amount of land to the generation of electricity? (It should be noted, however, that the land in between the turbines is not rendered totally useless and can still be utilized for agricultural purposes right up to the base of the towers.)

The environmental issues are very important, and strongly felt enough for the CEGB to judge that hilltop sites for wind turbines are 'environmentally unacceptable', even though none has yet been built on. Instead, interest has been focused on the development of lowland sites, where the wind speed is not so high, and on placing larger machines offshore.

## Out at sea

To position wind turbines out at sea certainly reduces the environmental risks and uncertainties. They would not be visually intrusive some miles from the shore, they would not take up agricultural land, nor would they create any noise or television interference. This policy would also have the added benefit of increasing the potential resource that is available. A very conservative estimate, taking into account such constraints as the proximity of the shoreline, sea bed conditions, depth of water, and other uses of the sea area, puts the annual energy available at about 240 terawatt hours. Other estimates, using different criteria, have put the resource much higher. One study carried out for the Department of Energy revealed that clusters of wind turbines built in shallow waters around the British coastline could produce 50 per cent more electricity than the present national demand. Since these turbines could also take advantage of the stronger and more consistent winds offshore, it is unlikely that shortage of this resource would be a limiting factor.

The construction, transmission and maintenance costs are higher than for land-based machines, and the technology involved in building them to resist the tough conditions at sea would tax the expertise of the companies who have made the oil and gas platforms for the North Sea. If they are to cope with wave heights of as much as 30 m (100 ft) in severe storms, as well as allow for the average depth of water and the geological characteristics of the sea bed, the offshore wind turbines will have to be very solid structures. In very shallow water an artificial island might have to be built, whereas in deeper water of around 40–50 m (130–165 ft), various methods of securing the structures would be used, from armoured concrete flat foundations to steel tripods or towers.

To make building wind turbines out at sea seem anything like economic sense, they would have to be built in a cluster or wind park; and each turbine has to be exactly positioned to obtain the best results. For as each one extracts energy from the wind, it disturbs the air flow for all those that are positioned downwind from it. So to allow the wind to regain its strength before running into another turbine, each machine has to be placed a certain distance from its neighbour. This distance is determined by the size of the rotor blades. It has been found that at a distance equal to five times the rotor diameter away from the turbine, the wind has reached 82 per cent of its undisturbed value; at ten times it reaches 92 per cent; and at twenty times, 97 per cent.

What plans are there to build these offshore wind parks? In the Netherlands there is talk of an offshore wind station of 200 machines, each one producing 3 megawatts, 50–80 km (30–50 miles) from the Dutch coast. In Britain, too, the CEGB, together with the Department of Energy and the Wind Energy Group, have looked in some detail at the problems of connecting a notional 2 gigawatt, 320-machine offshore array into the national electricity grid and have found that there were no 'insurmountable problems'. However, until more experience with land-based machines and land-based wind parks has been accumulated, it is unlikely that any offshore designs will be constructed. The most immediate prospect seems to be the first sea-based prototype which, it is hoped, will be constructed by the early 1990s.

The rather strange-looking oblique axis turbine

## The future

A few years ago, wind power was dismissed in Britain because achieving cost-effective electricity generation from wind turbines implied siting them on breezy hilltops. In England and Wales the relatively small number of suitable sites, combined with some possible environmental objections, made wind power a rather poor option. This view has, however, been altered by two major developments. Firstly, it has been shown that generators can be sited in clusters in shallow waters around the British coastline. This course is both technically feasible and has a very real chance of being economic. Secondly, tremendous advances have been occurring in wind energy technology. Using knowledge derived from the aerospace industry, costs have fallen while the ability of turbines to turn efficiently in lower wind speeds has increased. They have become more reliable and now offer some communities the prospect of cheap power. Even larger wind turbines, if produced in series, could be competitive with fossil-fuel plant – and some estimates have suggested that on sites with modest wind speeds they could compete with a single nuclear power plant.

Large megawatt-scale machines are still very much in the prototype stage and must be treated as such. They have significantly greater development costs than small or medium-sized machines, and are assembled from very much more expensive specially designed 'one-off' components. But if the development of these prototypes continues, there is every reason to believe that a commercial market will emerge in five to ten years from now.

For small and medium-sized wind turbines, a commercial market already exists. In Denmark, for example, 680 were bought in the period 1981–4. These smaller models can mostly be assembled from 'off the shelf' components, (i.e. the tower, gearbox, generator, etc.) although as with the larger machines, there is still some question concerning their reliability. But the refinements and improvements will happen as the technology expands and develops. These smaller machines together with megawatt-scale turbines could, it is claimed, offer commercial opportunities comparable with today's aerospace industry.

The emergence of wind power as a serious contender for the future of electricity generation could also have other effects that are more difficult to quantify. The impact of wind energy, like a number of other renewables, may result in a greater diversity of power sources, and in the long term may alter the structure of our electricity supply industry. It could become more decentralized, with wind machines deployed not only by the national generating boards, but also by individuals, small communities or local generating companies. It could also alter some aspects of the environment, with the North Sea, for example, not only studded with oil and gas platforms, but also with wind turbines.

The contrast between modern aircraft and their predecessors of 40 or 50 years ago is one that is now used by some wind energy promoters to illustrate the emergence of wind turbines. The old, traditional 'Dutch' windmill represented the state of the art over a century ago. Today's modern windmills are in their technological infancy. Already new designs and further innovative concepts have been put forward. If the pace of change proves to be as fast and as fruitful as in the aircraft industry, then wind energy may well prove to be the most promising means of intercepting solar energy economically, and on a scale large enough to make a substantial contribution to our present energy needs.

# Power from the Sea

### Turning the tides

If you have ever battled against the tide coming out while you are desperately trying to get ashore in a boat or when swimming, you will know just how strong tidal currents can be. Every day estuaries, bays and beaches are left looking naked and empty, their sand- and mudbanks revealed as tidal currents drain them of water. But what if we could harness the power of the tides, forcing them through turbines to generate power, providing a renewable energy source that causes no pollution and where fuel is free?

Tidal power is perfectly feasible and has been so for many years. Today's schemes involve huge barrages, with specially made concrete or steel caissons spanning estuaries, but the exploitation of tidal power has far humbler beginnings. As far back as ancient Egypt waterwheels were turned by the rise and fall of the tides, and in Europe their history can be traced back to about the eleventh century AD. One of the early English mills still exists at Woodbridge in Suffolk. The first was built there in 1170 and there have been a succession of them ever since.

The power of the Murray River in Australia harnessed to produce electricity. Hydro-electric plants are common now throughout the world, but even small rivers can be tamed to produce energy (see page 105)

The old London Bridge: one of the tidal mills can be seen on the left hand side (third arch from the bank)

The early mills were usually designed to grind corn and they operated by trapping the water at high tide behind a dam built across the mouth of a small coastal inlet. As the tide came in, the water would flow through the open gates of the dam and into the inlet. Once the tide had reached its height and the inlet was full, the gates were shut, trapping the water behind. When the tide turned, the imprisoned water was then forced to run back through the mill gates, this time turning a waterwheel and driving the mill machinery.

Other grander schemes were developed as time went on, like the one that was installed under the arches of the medieval London Bridge. Four water-wheels, each 6 m (20 ft) in diameter, were used not to grind corn but to pump water. They could be turned either by the rising or ebbing of the tide and lasted from 1580 until 1824, when they were dismantled.

Tide mills, like most other devices that relied on the elements, declined with the advent of steam power and by 1940 there were only 20 left in British Isles. However, the prospects of cheap and regular power that could be provided by the tides always kept research alive. The idea of a barrage across the mouth of the River Severn, for example, was first mooted in the middle of the nineteenth century and has re-emerged at various stages since. For the scale of the resource offered by tidal power is huge. It has been estimated, for instance, that tidal energy enters the English Channel from the Atlantic at a rate of 180,000 megawatts and that the greater part of this is wasted by friction at the bottom of the Channel. China, too, has an enormous tidal power potential of 110,000 megawatts – that is one-tenth of the world's total power needs – in some 500 or so bays along its coastlines. So how can the tides be exploited?

**Lunar power?:** One factor makes the job of harnessing the energy produced by the tides a lot easier: they are, very obligingly, regular as clockwork, their rise and fall following an entirely predictable pattern. Tides occur mainly because of the gravitational pull of the moon as it circles the earth, although they are also affected by the position of the sun in the sky and the rotation of the earth around its polar axis.

The earth gets two tides a day because of a rather complex series of gravitational and centrifugal forces. What happens is that the moon's gravitational pull induces water bulge on the side of the earth nearer to it, while on the far side, a second bulge is set off as a result of the centrifugal forces generated by the earth's rotation.

The range of the tides, however, is far from uniform and varies considerably from one place to another. Coastal regions or estuaries have a much greater range of tidal movement than the middle of the oceans, or seas, like the Mediterranean, that are protected from large current flows. The size of the tide is also affected by the relative position of the earth to the moon and the sun. When there is a full or new moon, all three are arranged in line. The forces produced are greater and the tides reach their maximum levels. They are at a minimum when there is a half moon, and the sun and the moon are pulling at right angles to one another.

**One cycle or two?:** The actual principle of turning tides into electricity-generating power plants has not really changed a great deal from the days of the early tide mills, although today's schemes rely on the expertise gained from hydro-electric installations. There are two ways in which the tide can be used. The simplest and cheapest involves generating power only when the tide begins to ebb. This means that only one cycle of the tide is being utilized. As the tide rises, the barrage sluice gates are left open, allowing the water to flow through into the rest of the bay or river, until the tide has reached its maximum. The sluice gates are then closed, trapping the sea water behind the barrage like a dam or river. As soon as the tide begins to ebb, the water can then be released through the turbines. This method means that the barrage could only operate for short periods of about 3 to 5 hours during each tide when there was a sufficient head of water. One way of increasing the power produced is to increase the head of water by pumping more water into the basin; this has improved the economics of the operation in some cases.

However, generating power from only one part of the sequence means deliberately losing half of the tide's energy as it comes in. A great deal of work therefore has been done on developing turbines that can work on both the emptying and the filling stages. This is a far more complex and costly process, and although the output of a barrage is certainly increased by this means, it is not doubled. Another way of getting more power from the same tide would be to create a second barrage. As a result there would not be just one basin but two, and by pumping water between them at the appropriate times, it would be possible to maintain a higher head of water and continue generating power for a longer time.

**Submerged turbines:** A rather less well-known technique for exploiting tidal currents is to submerge the turbines in the water. There is no head of water needed as the fast-flowing currents themselves turn the turbines. One advantage is that, whereas most tidal projects involve major investments over many years without any prospect of a return on the capital that has been put in, with submerged turbines the tidal current rotors can be built up gradually one by one into a large array, with each one able to generate power as soon as it has been completed. By positioning the turbines at various places along a coastline, each with different tidal regimes and operating at different times, it would be possible to provide a much more

POWER FROM THE SEA **101**

consistent baseload of power for any national grid. There would not be the large fluctuations in power characteristic of a one-cycle barrage, which can only generate power for a few hours twice a day.

However, for every advantage gained there is always another set of complications and tidal current rotors are no exception to this depressing rule. Problems of fishing nets, ships' anchors, marine fouling, and difficulties of securing and maintaining the rotors in position, all need to be resolved. There is also the question of the ability of the transmission system to cope with rotors that move very slowly under water when a high rotational speed is needed to produce electricity.

This concept of submerging rotors under the water has been suggested not only for tidal projects, but also for exploiting the energy contained in ocean or river currents. One rather extraordinary and very ambitious idea has been proposed which envisages underwater turbines used almost like windmills. Called the Coriolis-1 scheme, the idea is to place 230 turbines off the coast of Florida, near Miami, in the path of the Gulf Stream. The Gulf Stream, which flows at about 7 km/h (4½ mph) at this point, would drive the turbines to produce electricity. Each turbine is planned to be huge. The circular aluminium hull in which it would be encased would be 168 m (550 ft) across (that is the equivalent of three Boeing 747s placed wingtip to wingtip) and the 230 machines would produce an estimated 10,000 megawatts.

Not surprisingly, the project has its fair share of doubters and critics. Some argue that the scheme is not feasible, that the Gulf Stream would be slowed down by the massive turbines causing changes in the climate, and that they would damage fish stocks. The inventor, Professor Bill Mouton, and the project director, Peter Lissaman, disagree. They claim that at the most the temperatures would change by a couple of degrees, that the blades are turning too slowly to damage or kill fish, and that most of the technology involved is standard marine and offshore engineering. The money required to construct the 230-turbine array, however, is far from standard. The huge financial investment required may keep the project pinned to the drawing board. We shall see.

**Tidal power stations:** The prospect that tidal power stations would simply remain on the drawing board ended over 20 years ago with the construction of the 240 megawatt station at La Rance near St Malo in Brittany. The first large station of its kind in the world, it contains 24 reversible turbines so that it can generate power on both the ebb and flood tides. The turbines can also be used for pumping at high and low tide to increase the head of water. La Rance has been quite successful and has generally been used to replace the output from oil-fired power stations. It was quickly followed by a much smaller and far less ambitious project in the Soviet Union. This was an 800 kilowatt plant, seen as a pilot for later designs, which was built across the bay of Kislaya near Murmansk on the Barents Sea.

These two power stations signalled the start of a series of studies of other possible sites around the world. These began in Canada in 1967 where interest was focused on the Bay of Fundy, which has the largest tidal range in the world. Similar work followed in Australia, India, Korea, Britain (the Severn Barrage), Alaska and elsewhere in the United States, Brazil and Ireland. The only 'hardware' to emerge from all these investigations so far has been the completion of the Annapolis Royal pilot plant in Nova Scotia,

Canada, which was commissioned in April 1984 and some small developments in China.

**The Severn Barrage saga:** If any example is wanted of the different phases that tidal power technology has gone through over the last century to reach its present status, then the proposed Severn Barrage must take the prize. The British Isles have many bays and estuaries that have been suggested as suitable for tidal barrages. If they were all developed they could provide, according to the Central Electricity Generating Board, an annual output of about 4.5 terawatt hours per year (that is the equivalent of about one-fifth of the present British electricity demand). But the Severn Estuary, with one of the most impressive tidal ranges in the world, could produce nearly one-third of that total on its own.

The idea of a tidal power station on the Severn was first put forward around the middle of the nineteenth century, although it took until the 1920s for the suggestion to be officially reviewed. The Brabazon Committee, which took eight years to produce its report, concluded in 1933 that a barrage was technically feasible. It looked in depth at one proposal which would generate on the ebb tide and have a 800 megawatt capacity. Nothing came of it and interest ebbed and flowed like the tide itself until the mid-1960s, when another scheme was proposed by Professor Wilson of

**Some recently proposed tidal power schemes**

| Country | Site | Installed capacity (mW) | Annual energy (gWh) | Mean tital range (metres) |
|---------|------|-------------------------|---------------------|---------------------------|
| United Kingdom | Severn Estuary | 7200 | 13,000 | 9.3 |
| | Mersey Estuary | 525 | 1020 | 6.7 |
| | Strangford Lough | 210 | 530 | 3.1 |
| Irish Republic | Shannon Estuary | 318 | 715 | 3.8 |
| India | Gulf of Kutch: Kundla | 600 | 1600 | 5.2 |
| Korea | Carolina Bay | 480 | 1200 (approx) | 4.6 |
| Brazil | Bacariga | 30 | 55 | 4.1 |
| USA | Alaska: Knik Arm | 2000 | 5000 | 7.8 |
| Canada | Cumberland Basin | 1147 | 3420 | 10.5 |
| | Cobequid Bay | 4028 | 12,600 | 12.4 |
| | Annapolis Royal* | 20 | 50 | 6.7 |
| Australia | Secure Bay | 570 | 1650 | 6.2 |
| | Walcott Inlet | 1250 | 3940 | 6.0 |
| China | Jiangxia* | 3 | 11 | 6.0† |
| USSR | Lumbursky | 400 | ? | 6.0† |
| | Mezenskaya Guba | 10,000 | ? | 9.0† |

* Completed or under construction    † Approximate figures

OPPOSITE ABOVE LEFT The huge turbine at the tidal power station at La Rance, France

OPPOSITE ABOVE RIGHT An aerial view of the Annapolis Royal site under construction and as completed.

BELOW The plant's huge turbine before installation

Salford University. This would have produced four times as much electricity a year as the 1933 plan, and was sited further out towards the sea. Yet another idea was put forward in 1971, this time using two basins, until finally the whole 'to be or not to be' barrage came under official review again with the inception of the Severn Barrage Committee in 1978.

The Committee examined three schemes in depth: an outer barrage, which would generate on the ebb and produce about 20 terawatt hours of electricity a year; an inner barrage, which would also use the ebb tide and produce 13 terrawatt hours; and a staged scheme with one inner barrage, and a second one enclosing the whole of Bridgwater Bay to be built only when the first barrage was complete, and when a good economic case could be made. This second barrage would generate on the flood tide, allowing power to be produced four times a day. The Committee in the end recommended that the inner barrage proposal was the one on which attention should be focused and proclaimed it to be both technically feasible and economically sound. It would take 12 years to construct and would be a massive engineering project. When completed, it would generate the equivalent of 6 per cent of the United Kingdom's annual electricity needs at a cost of just over 3 pence per kilowatt hour.

The Severn Barrage would obviously have a considerable effect on the environment in the estuary, although this is the cause of some debate and disagreement among some of the scientists who have assessed it. It would create considerable disruption to the natural habitat during its construction, and its long-term effects could alter the character of the surrounding wetlands and mudflats, and as a result affect wildlife in the area.

But will it ever get built? The Committee reported in 1981 that the barrage was likely to be an economic investment, but that it was not as good a proposition as a nuclear plant. No decision has yet been taken to implement the proposal. The Central Electricity Board, who would be responsible for accepting its power, is not enthusiastic. 'Although it is possible', the Board states, 'that the government might decide on policy grounds that a Barrage should be built, there is no incentive for the CEGB to seek to build it in preference to nuclear power stations.'

But is it just a question of economics and comparisons with nuclear power stations in the light of the current economic analysis that prevent it from being constructed?

Professor Wilson, a long-time expert on tidal power, thinks the present view is misguided. He equates tidal projects with hydroelectric schemes and argues that the economics will appear very differently ten years after construction. 'How many remember now', he says, 'how the hydroelectric schemes of the North of Scotland Hydro-Electric Board were criticized at the time? Mostly built over 20 years ago, they have an indefinite life ahead of them, producing energy at the lowest cost in Britain.' He continues:

> What, one wonders, will our children and grandchilden say? On the one hand they will have two worn-out reactors on permanently sterilized plots of this small country, to be guarded indefinitely, with toxic waste to be managed for generations. On the other hand, they would have a benign power scheme, with an indefinite life ahead, producing 13,000 million kilowatt hours per year. I wonder will they agree with the Severn Barrage Committee that the nuclear plant was the better investment? (Professor Wilson in a paper to the Fourth International Conference on Energy Options, April 1984)

### Hydroelectricity

The contribution that hydroelectric plants have made over the last 20 to 30 years has borne out Professor Wilson's remarks. For the days of the water-wheel beside the local stream have given way to the widespread development of large generating stations all over the world. Huge dams and reservoirs have been built to exploit the power in rivers and waterfalls to drive a turbine. Today hydroelectric stations supply nearly a quarter of the world's electricity, and the contribution will continue to grow. Some countries already receive all their requirements from large river schemes, and others get at least two-thirds of their electricity from this renewable energy source.

The potential for more schemes is still huge. Whereas Europe is already exploiting over 50 per cent of its possible hydroelectric sites, the Third World only uses about 8 per cent of its estimated capacity. One survey suggested that between 2 and 6 per cent of Africa's and Latin America's hydro-power resources have been exloited.

However, developing this potential is not easy. The projects, like tidal stations, involve massive capital costs and take years to reach fruition. There are also considerable ecological restraints, as the dams can involve the flooding of farmland, a raising of the water table and an increase in waterborne diseases. Local wildlife can often be severely affected.

Yet the development of new projects still continues. Quebec, for example, completed Phase I a few years ago of a hydro scheme of gargantuan proportions. It will eventually consist of 9 dams and 170 dikes, creating a reservoir the size of the state of Connecticut in the United States to provide a massive 11,400 megawatts. Phase II, which is due to be ready by the 1990s will bring *La Grande Complex* near to its completion. But such huge undertakings – this must be one of the largest engineering projects of the century in terms of sheer size – will not happen everywhere and there are tremendous resources still available in much smaller rivers, with comparatively low flows and heads of water.

**Mini-hydros:** The potential power to be gained from even these small sites is considerable. Even in the United Kingdom, a country not noted for its large hydroelectric reserves, 560 sites were identified in Wales alone that were suitable for mini-hydro plants. If all of them were developed, according to a Salford University study, they would produce enough electricity to save at least the equivalent of 80,000 metric tons of oil a year. Other countries have similar findings: France has between 3000 and 4000 sites; Sweden 1300, and the United States could generate about 11–13 per cent of its electricity. The environmental impact for these mini-schemes is less than large-scale projects and they can provide power on a decentralized basis. But even if Britain did not develop its mini-hydro potential, there is a much greater source of energy waiting to be tapped.

### Wave power

In the seas that surround Britain lies a source of wealth which, if it could be gathered, could prove to be as valuable and as useful as North Sea oil. What Britain, like many other countries, has is pure energy, and unlike oil this resource will last for ever. It is the energy in the waves themselves and if we could harness it we would have gone a long way towards providing the power we shall need in the twenty-first century. For wave energy, it has been estimated, could supply up to a third of our electricity needs.

   This potential powerhouse in the sea has been taxing the brains of many researchers and no fewer than 340 patents have been filed in Britain alone for wave-power devices. However, the first recorded patent was not in Britain but in France. The Girards, father and son, filed it in Paris on 21 July 1799. Their idea was to build a gigantic lever, whose fulcrum was to be located on land, while the 'pick up' end to extract the wave energy would be secured to a pontoon which would rise and fall with the waves.

ABOVE One of the most impressive hydro-electric schemes in the world: the Hoover Dam on the Colorado River in Arizona

LEFT There is still a tremendous potential for small hydro-electric schemes around the world. This waterfall is in Kenya

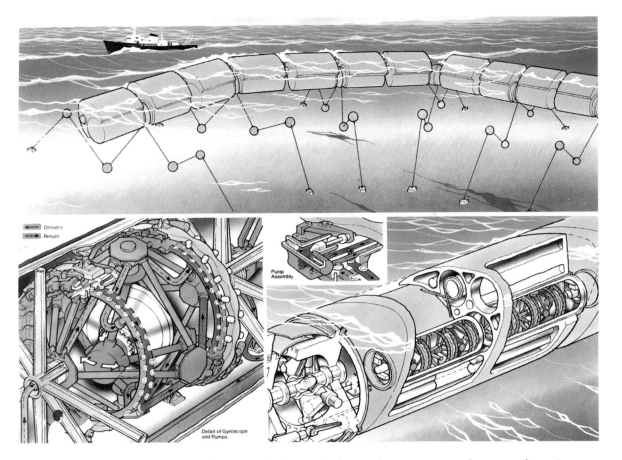

Delivery
Return

Pump
Assembly

Detail of Gyroscope
and Pumps

The Edinburgh Duck,
showing the gyroscope and
pumps in detail

The first recorded practical use of wave power also came from France, this time at Royan, near Bordeaux in 1910. A vertical borehole in a cliff was used by Monsieur Bochaux-Praceique to drive an air turbine and supply his house with one kilowatt of power. It was a natural oscillating water column. This design is now being used for a large-scale device which, it is hoped, will become Britain's first ever wave-power staton. There have been many other designs, too, all trying to grapple with the problem of converting the low, random and oscillating motion of the waves into the high-speed unidirectional rotation required to generate electricity.

This is not an easy problem to solve. If the device is to be effective and absorb most of the energy in the waves, it will have to be efficient. It will have to cope with waves of varying height, frequencies and directions, and weather conditions of all kinds, as well as operating reliably with a design life of at least 20 to 30 years. At the same time it must not be over-protected or over-engineered as it would then prove too expensive, or consume more energy to make than it would extract in its entire working life. So what are the main contenders?

**Salter's Ducks:** In the early 1970s Stephen Salter of Edinburgh University developed and refined a series of 'ducks' which nodded up and down in the path of the waves, and in doing so extracted some of the energy from them. What he had first envisaged as the most obvious means of extracting energy from the sea 'was something like a lavatory ball-cock, bobbing up and down'. But when he tried it, he discovered that it only extracted 15 per cent of the energy in the wave.

Salter's Duck is far more efficient. As the wave slaps against each unit, it moves up much like a body of water, and as it moves, it accepts the energy of the wave that hits it. And because the Duck mimics the wave, it is also capable of passing on that energy to drive a toothed belt, which in turn powers a hydraulic pump. That pump is fed with hydraulic fluid and the continual up-and-down movement of the Duck forces the fluid back out again under pressure along a central spine which links all the units together. The hydraulic fluid from all of these is then pumped into a central power pack, which by means of a motor extracts all the energy from the fluid and uses it to drive a generator to produce electricity.

However, this version of the Duck ran into serious engineering problems and Salter had to think again. This time he proposed an extremely novel solution, based on gyroscopes. The second version has two pairs of gyroscopes mounted in a cannister. The tossing motion now causes the entire gyro assembly to rotate as it tries to resist the movement of the waves, the gyros precess and recess and this in turn drives the hydraulic pumps. The use of gyroscopes has a number of attractions: they can store vast amounts of energy providing a kind of 'spinning reserve'; they reduce the strain on the spine; and they can be hermetically sealed in the factory, so that no seawater can seep into the delicate mechanism. Salter hopes that this would make the device maintenance-free for its lifetime.

The Kaimei which will be tested at sea in 1985 with a modified hull

The Duck is a very complex and innovative piece of engineering and is the most technically advanced wave-energy system. At the moment there are a number of problems to be ironed out and the project is still at its research and development stage. But if all the difficulties can be solved, then the Duck should have an optimistic future ahead.

**The oscillating water column:** While the Duck uses advanced hydraulics and converts wave energy mechanically to produce power, the oscillating water column (OWC) operates pneumatically. It was pioneered in Japan by Yashio Masuda and has already been turned to good use, powering more than 300 navigational buoys in the Sea of Japan. The device itself is based on the idea of an inverted can, half submerged in the water, and works on the principle that the waves regularly compress a column of air trapped at the top to drive a turbine. As the wave approaches, the pressure pulses assocated with it set the column of water oscillating up and down in time with the period of the waves. The whole column then acts as a piston: as it rises the air at the top is compressed and forced up and out through an overhead valve and into a low-pressure air turbine. As the wave trough passes the pressure drops and a system of simple valves ensures that the air is sucked back into the chamber via the same turbines. And so it repeats itself *ad infinitum*.

This concept has been exploited by several designers in Britain. Masuda himself has developed a wave-power machine that resembles a ship, but has large holes intentionally cut into the hull at water level. It is called the *Kaimei* and has 22 individual compartments, each of which traps a water column in the same way as the buoy. It is a full-scale prototype 80 m (262 ft) long, 12 m (39 ft) wide and 7.5 m (25 ft) deep, but during its tests in the Sea of Japan its performance proved disappointing.

**NEL breakwater:** The OWC idea has been taken one stage further in Britain and after extensive research and development looks likely to become the

country's first wave-power station. Using the OWC principle, the National Engineering Laboratory has proposed a design that incorporates four OWC units and will generate 6.4 million kilowatt hours annually. The whole design will be 84 m (276 ft) long, but because each column has its own individual set of generating plant, more can be added alongside to expand the capacity of the power station in the future.

Moored near the Isle of Lewis in the Outer Hebrides, the first full-scale prototype is planned to be built in 14 m (46 ft) of water in a steeply shelving site, where the annual wave energy available is around 30 kilowatts per metre. The plant will generate 5 megawatts of electricity, expected to be at a cost to the customer of about 6 pence per kilowatt hour. If all goes well the team hopes that by 1987 Lewis will have its first fully operational wave-power station, replacing the existing expensive diesel generators.

**The Belfast buoy:** Although the breakwater design turned its back on a moored floating device to resist the continual battering it would receive from the waves, research at Queen's University, Belfast, Northern Ireland, remained faithful to this principle. A team, under the direction of Dr Wells, began in 1975 with the development of an air turbine with special qualities. For instead of a device able to accept air from one direction only like most turbines, Dr Wells devised a rotor that would accept air from either side and still turn in the same direction. It was an invention that was turned to good use, for the team developed a wave-powered navigation buoy, and a design for a floating 2 gigawatt station in a project commissioned by the UK Department of Energy.

However, the Belfast design, like a number of others (those from Lancaster and Bristol Universities, for example), was refused further funding by the Department of Energy in a major review on wave power carried out in 1982. Although the Belfast team has not given up its work, its attention is now being focused on small-scale developments like the buoy rather than on large power stations. Another design to suffer a similar fate was the SEA-Clam.

**The Clam** The Clam works by using a series of flexible air bags to extract energy from the waves. Each unit consists of a long, hollow, floating space made of reinforced concrete, which acts as the main structural element of the device. Along this spine is mounted a line of submerged, specially reinforced rubber air bags. The bags 'breathe' air in and out in response to the waves, and generate electricity by driving the air through the turbine. A 25 m (39 ft) long, 12 metric ton model was successfully built and tested on Loch Ness in Scotland during 1981 and 1982.

The Clam programme was always aimed at designing and evaluating the prospects for the Department of Energy's huge 2 gigawatt station. It was planned to consist of 240 separate units. Each one would have 10 flexible air bags, driving a single self-rectifying turbine, and the whole system would be able to move around the single anchor point automatically to adjust to any change in the direction of the oncoming waves.

However, as a result of the Department of Energy's 1982 review, which stated that wave power did not 'appear sufficiently attractive to justify the commitment of a large sum of money to a major sea trial', the next stage, the development of a 10 megawatt prototype, was dropped. But rather than abandon their efforts, the group from Lanchester Polytechnic, Coventry,

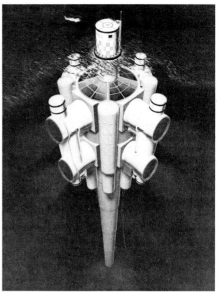

under Dr Norman Bellamy decided to go ahead privately with much smaller versions. They are hoping to build and market wave devices in the 250–1000 kilowatt range, aimed at providing economic power for remote islands and coastal communities.

ABOVE LEFT An artist's impression of the NEL Breakwater. The electrical power is transmitted by a conventional submarine cable to the shore
ABOVE RIGHT An artist's impression of one large OTEC station with details of its workings (see page 112)

**The United Kingdom programme:** In 1976 wave energy had been considered to be 'the most attractive of the renewable resources' and was given a high priority by the Department of Energy. It was envisaged that by the year 2000, wave energy could possibly contribute up to 15 per cent of the present electricity demand, and by 2025 it could have reached 50 per cent. In 1984 the British wave-energy programme lies buried in a graveyard of technological has-beens. A very 'limited programme of research to explore possible improvements in the technology' is being undertaken, but any wave-power projects that are continuing are having to look elsewhere than to the government for funds. Perhaps the real damage was the lack of faith that wave power could contribute substantially to our energy needs, just when the real investment was needed.

The Lewis project is encouraging (and at the time of writing it looks as if another breakwater device will be built, also at an island site) but it would be a tragedy if the country were to lose not only the expertise that has made Britain a world leader in this field of research but, more importantly, the chance of developing a source of energy that is inexhaustible. Wave power consumed £15 million up to 1982: that was the total cost of its research and development. If such a relatively low level of investment had been imposed on nuclear power development in the 1950s, and had continued to apply to the fast breeder and fusion projects, then nuclear power would not have been able to make the important impact it has on the energy industry. If the same criteria had been applied by a government review body to nuclear research when it was in its infancy as has been applied to wave power, it would have been interesting to note what would have been said.

The view that wave power is unattractive economically is one that is not universally accepted either in Britain or abroad. It is claimed that new concepts are emerging which, together with a more complete optimization study, would reduce costs by a significant factor. One of these new concepts

The Clam on test in Loch Ness, Scotland. This is a one-sixth scale model of a new 650 kw device, designed with small island communities in mind. The two artist's impressions ABOVE and OPPOSITE show the proposed completed unit, with six bags and six Wells turbines

is to add two parallel projections – in effect a harbour – in front of a wave-energy device. This has the effect of introducing harbour resonances, resulting in an increase in the amount of energy a device can capture in waves that are irregular and coming from all directions. Early theoretical and experimental results have been obtained from an oscillating water column, both with and without such a harbour, and the projections have improved the performance significantly (one report showed that they doubled the performance of the device).

**Waves from Norway:** Although Britain has not proceeded with developing a full-scale plant, other countries have gone ahead. Norway, which has done some of the research into the harbour concept, is building a small power station that will operate on the oscillating water column principle. The power station itself will be built into the bed rock of Norway's west coast. A specially built shaft, blasted out of the rock, leads up from the sea to a turbine. As the waves make contact, they will have a 'piston-like' effect on the air inside the shaft, increasing the pressure and forcing it up and through the turbine. It will generate about 1 million kilowatt hours per year, which is enough to supply the needs of between 30 and 40 homes.

The Norwegians have been working on wave power for some years and have developed a buoy that looks promising enough to be scaled up into a power plant. They have found that when a buoy is timed to the period of the waves, it can extract more energy than when it bobs freely up and down. So, sensors have been mounted on the buoy which detect the pressure of incident waves. This information enables an on-board computer to determine when the buoy should carry out its oscillations and get the best out of the wave. It is helped by a vertical shaft which runs through the centre of the buoy and is fixed to the sea bed.

The design has been tested at the model stage and plans have been drawn up for a 200 megawatt wave-power station, consisting of between 400 and 500 buoys, deployed in strings of five, along 10–15 km (6–9 miles) of coastline.

Whether this station will ever be built is another question. For if a scheme that was put forward in Mauritius is anything to go by, it does not stand too good a chance. A proposal for a wave-power plant there has been on the drawing board for 20 years. It involves building a 5 km (3 mile) long sloping wall across Riambal Bay using the coral reef in the bay as bases. This wall, with a crest height of 2.3 m (7½ ft), would be linked to the shore by side walls at each end, creating an enclosed reservoir. Waves crashing over the sloping wall would fill the reservoir to a height above sea level. The water would then be pumped into a higher level reservoir, supplying a conventional hydroelectric plant, which would feed the island's public electric system. If the system does get the go ahead, a pilot wall about 1 km (1100 yd) long will be built. But for countries like Mauritius and the other Indian Ocean states, there may be another potential renewable source that could offer very economical power indeed.

### Exploiting the oceans

Tropical and subtropical islands may have many blessings, but unfortunately a cheap form of energy does not seem to be one of them. Yet if present developments continue, the warm surface water of the oceans surrounding these island paradises could offer an inexhaustible source of renewable energy.

The idea is to exploit the large temperature differences that exist between the warm surface water of the ocean and the cold deep water at around 1000 m (about 3300 ft). The warm surface water in tropical areas is often between 25° and 30°C, whereas the water at depths of 500–1000 m (1650–3300 ft) has flowed from the polar regions – predominantly the Antarctic – and temperatures can fall to 4°–5°C. The concept is called Ocean Thermal Energy Conversion, or OTEC for short, and it uses the 20° difference in a conventional heat engine cycle to generate electricity.

OTEC owes its origins to a Frenchman called Jacques Arsenè d'Arsonval, one of whose electrical devices, in 1881, was rather like a common domestic refrigerator, but with the input and output terminals reversed. However, 50 years had to elapse before another Frenchman, Georges Claude, attempted to turn the OTEC concept into practical use in Cuba and off the coast of Brazil in the 1930s. Since then the idea has come and gone in the scientific community, remaining an attractive energy source in theory but never achieved in practice. But now seven national OTEC programmes have made substantial progress and for a number of locations the economics of the system are promising.

There are two main ways in which the ocean's thermal gradients can be harnessed. The first is an open-cycle process, originally used by Claude in his early experiments. The warm surface seawater is flash-evaporated under vacuum to produce low-pressure steam. This steam drives a large turbine and electricity is produced. The steam then has to be condensed and this can be done either by direct spray contact with deep, cold seawater, or in a heat exchanger.

Although it is more expensive, the second method has a very big advantage: the water that is produced is fresh, as the seawater salts have been removed during the initial flash-evaporation process. A 5 megawatt plant, for example, would provide 90,000 litres (24,000 US gal) a day. On many tropical islands, where fresh water can be as highly prized as electricity, an automatic desalination plant coupled with electricity production would be a valuable asset.

This latter method of condensing uses a closed cycle and seems to be the favourite contender for a series of 'first generation' OTEC plants. In this cycle the warm surface water is pumped through a heat exchanger to vaporize a secondary working fluid with a low boiling point like ammonia. The vapour that is produced expands to drive a turbine, generating power in the normal way. To complete the cycle, the spent vapour is passed through a second heat exchanger, where it is condensed by the cold water extracted and pumped from the ocean depths. The secondary fluid is then pumped back to be ready for the cycle to begin again.

OTEC plants are not very efficient – around 2.5 per cent with a 20°C temperature difference – especially when they are compared with a typical figure of 30 per cent for a fossil fuel power station. Although this is very low efficiency rate by modern standards, there are other advantages.

One is that the deep cold water could help the local fishing industries. There are much higher concentrations of nutrients in the deeper ocean layers than in the waters nearer the surface and these would be brought to the surface with the water used during the OTEC condenser stage. The nutrient-rich cold water could then be used directly for the production of shellfish. Another advantage is that the electricity produced could be used *in situ* to supply energy-intensive products as OTEC plants can either be built on land, or they can be floating installations moored some distance from the shore.

If the plant is located onshore it has a long underwater pipeline stretching out to reach the constant cold deepwater it needs. Designs for OTEC stations in Tahiti, Jamaica, Hawaii, and on India's Andoman and Nicobar and Lakshadweep islands, and the Indonesian island of Bali, have been drawn up and a number of detailed experiments have been done. The Bali scheme is awaiting a decision at the time of writing. If it goes ahead, it could be the first working unit of a size large enough to demonstrate the viability of the technology.

Designed to feed the local grid, the land-based plant on the north coast of Bali would extract water from a depth of 550 m (1800 ft) at a site 1.5 km (1640 yd) offshore. The plant, which is seen as a prototype, will involve placing a 1.8 km (2000 yd) long, high-density polyethylene pipeline to carry the cold water to the land-based site and will produce less than one megawatt of electricity.

A different solution to the problem of carrying the cold water has been proposed for the South Pacific island of Nauru, where a Japanese consortium is finalizing designs for a 2–2.5 megawatt plant, fed by twin bore tunnels. Nauru, with ocean temperatures as high as 30°C and a temperature difference of 24°C, provides an excellent site for an OTEC plant. A small test plant, rated at 100 kilowatts, was already built there in 1981. It used freon as the working fluid in a closed-cycle process and was operational between 1981 and '82. The plant was a success, producing more power than was expected and no major mechanical problems developed during its 12-month life. However, a severe tropical storm at the very end of the test period created intense wave vibration and caused the cold water pipeline to rupture. So the team decided to abandon the pipeline techniques and a tunnel scheme has been chosen to avoid another failure of this kind.

The Japanese have also developed a station of their own, albeit a very small one. Based at Tokuno Shima, one of the southerly islands close to Okinawa, the plant uses a very sensible way of boosting the water tempera-

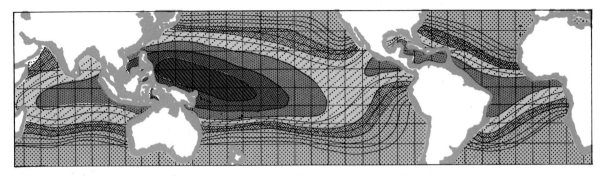

ture up to tropical levels. It utilizes the waste heat from a conventional power station, increasing the efficiency of the OTEC unit. The Japanese plan to develop a viable land-based scheme from this unit and then try to establish a floating plant at a later date. At present they are hoping that they will have a 10 megawatt unit open before 1990.

The United States is also pressing ahead with a much larger scheme in Hawaii, although American plans are not what they used to be. In 1980 the technology received a considerable helping hand in the United States with the passage under President Carter of the OTEC Research and Development Act, which established a demonstration fund, backed up by $2 billion in loan guarantees. With the arrival of the Reagan administration, funding was cut back and the idea of peppering the Gulf of Mexico with OTEC stations to provide base-load electricity around the clock has now become a distant dream. However, the Americans are working on a 40 megawatt land-based demonstration plant in Hawaii. The first phase was completed in 1983 and phase two is under way. Plans have also been drawn up for other sites in Puerto Rico, Guam, the northern Marianas and the US Virgin Islands.

What, then, are the prospects for OTEC power? The way to further progress lies in the construction of representative scale plants. The gap between the relatively small-scale tests carried out so far and the fully economic plants (about 5–8 megawatts on land and 25–40 megawatts for floating units) is one that will have to be bridged over the next decade. If all goes well, then it might be conceivable that by the turn of the century OTEC plants comparable to established generating systems will be feasible. By then, too, some of OTEC's environmental effects of cooling the oceans, even though on a small local scale, will be apparent. But while one group of researchers attempts to plumb the world's oceans, another is attempting to plumb the earth's hot core in the pursuit of geothermal power.

ABOVE At the best sites for OTEC plants, coloured red on the map, the surface water is 23°C warmer than the water 900 m (3000 feet) below. The plants could also use slightly cooler waters, shown in shades of orange where the temperature difference is usually at least 20°C

The small 100 kw pilot OTEC plant on the West Pacific island of Nauru. The power generaton unit is on the left side: the pipelines for cold and warm seawater are on the lagoon. Cold seawater is pumped up from a depth of 580 m (1900 ft)

# CHAPTER EIGHT

# Geothermal Energy

Since the famous eruption of Vesuvius destroyed the towns of Pompeii and Herculaneum in AD 79, there have been many violent volcanic eruptions that have brought widespread catastrophe and loss of life. The destructive power of a volcanic eruption releases tremendous amounts of energy. Whole landscapes can be changed overnight, with lava pouring out of the earth at temperatures varying between 600° and 1500°C. Compared to the temperatures of the earth's core – estimated at 4000°C – an eruption may be only a blister breaking on the earth's skin, but volcanoes are good indications of an alternative energy source which could provide every country in the world with a contribution to its heating or electricity needs.

The resource is geothermal energy. It is literally the heat of the earth, and it is one of the few 'alternative' energy sources that does not rely on the weather or the tides to work. It relies instead on the earth's hot magma and the natural decomposition and fission of uranium, thorium and potassium isotopes for its heat: for the earth is rather like a very weak natural nuclear reactor that has been switched on for 250 million years.

At the moment we exploit geothermal energy through its ability to heat fluids trapped in the earth, where the fluid is simply the means of

The 'Old Faithful' geyser at the Yellowstone National Park, Wyoming

transporting the energy. In volcanic areas or the tectonic regions where the earth's crustal plates meet, deep-seated cracks occur enabling molten rock to reach almost to the surface. This heats the naturally occurring ground waters to high temperatures, and they sometimes burst through fissures to reveal themselves as geysers or hot springs. However, more often than not, these aquifers of hot water are trapped within a layer of permeable rock several kilometres below the surface. To harness their energy, wells, similar to those of the oil and gas industries, have to be drilled.

## Drilling for heat

An ideal geothermal well is hot enough to produce what is called 'dry steam', with no liquid at all. Unfortunately, there are not too many of them and only about 5 per cent of the earth's geothermal reserves can be brought to the surface in this way. More commonly, the reserves arrive at the well head as hot water of varying temperatures – from 30°C upwards. How that geothermal fluid is then used depends on its temperature and its mineral content.

But why should the mineral content affect the way the geothermal fluid is used for energy? The problem is that most fluids are 'dirty'. They do not emerge from the ground so pure that you could pour them into your coffee cup: instead some carry up to 35 per cent of 'total dissolved salts'. These minerals in solution create very large headaches for any engineer, because the damage they can cause can render a power station useless. For unless the engineering and construction methods are just right, a 76 mm (3 in) pipe carrying geothermal fluid can be constricted to 13 mm (½ in) diameter flow in just 100 hours of operation.

Early attempts to exploit the geysers in the Mayacanas Mountains in California during the 1920s foundered largely because of the corrosiveness of the steam, which destroyed pipes and turbines. It was not until the appearance of stainless-steel alloys that producing electricity from geothermal fields began to make an impact. Today the area produces enough electricity to supply the needs of about one million people, and new applied engineering techniques are now making it possible to extract the potential from these fluids and still control the problem of corrosion and scaling.

## The flash cycle

High-temperature water is being used around the world to generate power by what is referred to as the 'flash cycle'. To produce electricity from geothermal resources that are predominantly liquid, the same method is used. Basically, hot water from the subterranean reservoir, thousands of feet down, is pumped up to the surface under pressure. It is channelled into vessels where the fluid is allowed to boil, or flash, to produce steam. The steam, about 20 per cent of the fluid, is then fed through a standard steam turbine in the conventional manner.

The flash cycle works by reducing the pressure of the fluid in the vessel in order to get it to boil. But that pressure reduction cannot always be controlled. Just bringing the fluid to the surface can itself cause flashing. As a result the fluid can arrive at the well head as a two-phase mixture of steam and water: and until recently only the steam could be used to generate electricity.

A new turbine developed in Utah in the United States could change that. It works by pumping the steamy geothermal liquid through a nozzle. As it emerges the pressure is dropped, producing more steam and giving the water more energy at the same time. At that point the fluid is captured in a rotating drum where the effect of centrifugal force separates the steam from the water. The steam is routed off to a normal steam turbine and the water is propelled to a special liquid turbine to produce yet more electricity.

The early tests have been completed and a full-scale trial is now under way on a 20 megawatt plant. If that proves to be successful, three 50 megawatt designs will be built by 1990. But while more can be made of the resource, what about the corrosion?

### Stopping the rot

One of the problems for the engineer is that geothermal reservoirs vary considerably in their mineral content, and a technique that might be suitable for one field may be totally unnecessary in another. However, various attempts at finding methods of dealing with the high concentration of salts and sulphides are under way, with most of the work being carried out in California.

Two pilot plants, each rated at 10 megawatts, are investigating different systems. The first alternative, known as the Brawley scheme, has already been changed. The original concept had been to keep everything in solution by maintaining a high pressure and so a high temperature. However, problems developed after the flashing process, as the temperature drops and the minerals begin to turn to solids. Instead, the fluid is now discharged into cooling ponds to allow the minerals and salts to drop out of solution and be taken away. The fluid can then be returned to the earth.

The second plant, at Salton Sea, also in California, treats the briny water thoroughly and removes the solids. Its engineers have discovered a way to insert a crystallization system in the fluid line before the brine reaches the flash points. The fluid passes through a series of filters and the minerals are taken off before they can reach any of the plant's machinery to cause damage.

The technology chosen for the Heber Reservoir, just south of the Brawley and Salton Sea projects, is very different as the brine has a much lower salinity than the other two fields. It uses a process called the 'binary cycle', and the world's first full-scale commercial facility is in the process of being built here. It is hoped that the binary cycle will not only minimize the problems posed by the brine, but will also offer a technique for exploiting reservoirs that have more moderate temperatures, where the flash cycle is effectively ruled out. The system uses heat exchanges to transfer the heat from the hot geothermal fluid to another liquid. This second liquid, known as the working fluid, is specially selected so that it has a boiling point equal to the prevailing temperatures of the hot geothermal brine. So as the heat is transferred, the working fluid vaporizes and is used to drive a turbine.

The brine that is left still has to be disposed of, whatever process is used. The main method is to re-inject it into the aquifer via another borehole right on the edge of the reservoir. It cannot be close to the well as otherwise the cooled liquid would simply ruin the rest of the reservoir. By re-injecting some distance away it is hoped that the brine will be geologically reheated, prolonging the life of the reservoir. As for the minerals, some of them could be harnessed. In fact the first power station to trap geothermal energy, built at Lorderello in northern Italy in 1904, was originally designed to produce boric acid as well. A number of other valuable minerals could be exploited in the same manner.

However useful these extra resources might be, they can only really be applied to electricity-generating plants. If the hot aquifer is to be used for heat, it is hardly possible to pump the corrosive brine around the district heating circuit. Instead the geothermal heat is managed in a very similar way to the binary cycle described earlier, but this time the working fluid is water, which can be pumped around the homes without a problem. Resolving these difficulties has been the great challenge of the last 20 years and work will continue. It has been instrumental in taking geothermal power to today's position where it provides over 3500 megawatts of electricity and an even larger figure in terms of heat: and the figures should keep on rising.

### The world resource

The inhabitants of settlements like Boise in Idaho in the United States would be rather amused to learn that geothermal power was a 'new alternative technology'. They are using the same hot water wells that their predecessors first tapped in the 1830s. So too would the people of Iceland. Their first hot water borehole was drilled in 1928 at Thottalaugar, a hot spring traditionally used by the people of Reykjavík for doing their laundry. The borehole was used to heat 2 schools, 70 houses and a swimming pool. Today Iceland has made good use of its geothermal sources and 70 per cent of its homes are heated by them.

This expansion is not confined to countries like Iceland. Energy from the hot water or steam trapped undergound in fault zones is being developed

Tapping the earth's geothermal resources at the Geysers field in California

One of the geothermal energy projects under way in remote areas. This one is at Kemokan in West Java

along the Pacific coast of the Americas, the Rift Valley of East Africa, around the Mediterranean Sea, and along the western side of the Pacific Ocean. Some of the most prolific and cheapest sites are in highly volcanic areas where the heat is near to the surface – Mexico, Costa Rica, El Salvador, Nicaragua, the Philippines, Indonesia, Kenya and Ethiopia – and development in these countries is moving fast.

The Philippines is one of the world's foremost users of geothermal energy and has a potential reserve of 10,000 megawatts. At the moment it has only just over 600 megawatts on stream, but there are plans to more than double this figure by the end of the century. Indonesia, too, is stepping up its development and exploration following the successful commissioning of the country's first plant at Kamojang in Java. The small 30 megawatt station is the beginning of the country's drive to install 100 megawatts within the next ten years. In the same Pacific area, which stretches from New Zealand through the archipelagos of Indonesia and the Philippines, to China, Korea and Japan, similar plans are under way.

The United States might have another huge geothermal reserve at Beowave in northern Nevada to complement its Californian schemes. The first stage of the development, it is hoped, will be complete by the end of 1985. The geothermal resources here, covering large portions of sparsely populated areas, could yield the energy equivalent of a giant oilfield, providing enough electrical power to meet the needs of 180,000 people.

Nevada may be sparsely populated around the Beowave field, but Europe's geothermal resources often occur in quite densely peopled areas. Most European projects have been large, because the temperature of most of the aquifers have required fairly sophisticated technologies and they have been aimed at providing heating for houses or agricultural users. Hungary, for example, has drilled hundreds of deep wells in the Panonian basin, and

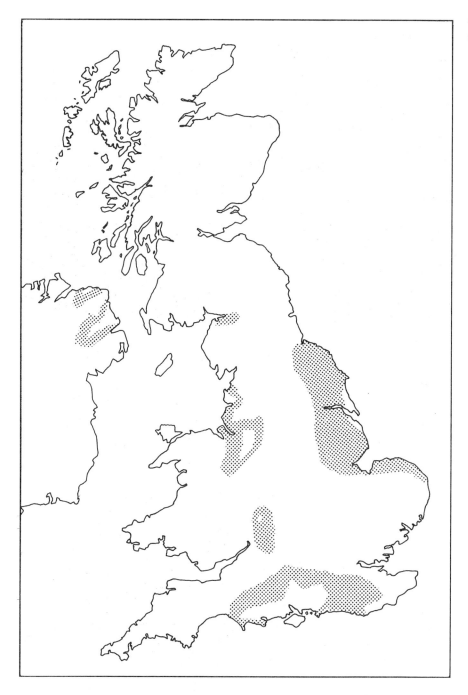

The possible geothermal sites in the British Isles

France had 23 schemes operating by the end of 1982. Over 100,000 homes in the Paris suburban area are heated from aquifers and by the 1990s it has been estimated that France will have an installed geothermal heating capacity equivalent to an annual saving of 10 thousand million metric tons of oil.

In Britain there have been few attempts so far by engineers to drill for heat, although there are seven large sedimentary basins which are thought

Mining the Earth's heart: a
geothermal scheme for a
block of flats at Villeneuve-
la-Garenne in France

Map showing European
sites suitable for
exploitation of geothermal
energy

to contain water at accessible depths, and with temperatures ranging between 40° and 100°C. The basins in Northern Ireland, the Wessex area, East Yorkshire, Lincolnshire and Cheshire are the most favoured for development and they could possibly supply a significant proportion of the space and water heating needs of the areas. A study by the Institute of Ecological Sciences calculated the total identified resource to be equivalent to $2 \times 10^9$, or 460 million tons of coal.

It is certainly a large resource, but it is one to which little attention has been paid in the past. However, there are more promising signs now that our geothermal supplies might well be exploited for heating. A long test took place in Southampton to trap a small part of the Wessex basin. Coupled to the idea of a new development in the City Centre there, the study showed that the aquifer would be able to provide heat far more cheaply than any of the competing conventional methods and would have a 20-year life. However, subsequent tests on the well found that it was not as large as was originally thought, and Britain's first demonstration of the use of geothermal energy has had to be scaled down already. Now it is planned to provide heat for the existing Southampton Civic Centre and the Central Swimming Baths.

Another test borehole is being drilled in the Grimsby-Cleethorpes area and there is also one in South East London. These wells, however, are exploratory and are a long way off from turning into geothermal heating schemes. But while British efforts to exploit aquifers have so far made little headway, a team at Camborne in Cornwall has been exciting a great deal of interest, working on a very different idea.

## Hot dry rocks

The idea, in principle, is very simple. Anywhere on earth the temperature increases as you go deeper towards its core and it is quite likely that the rock will be non-porous. So if you could develop a process for producing fractures and flow paths through the rock, it would be possible to extract the heat in the surrounding rock by putting cold water in at one end and getting hot water out at the other. Instead of having a natural aquifer, the hot dry rock concept creates a man-made geothermal reservoir and, by drilling, an artificial geyser. This means that, in contrast to natural reserves, any site in the world offers a potential location. The only constraint is how deep you can drill.

One of the earliest references to the idea of extracting heat in this way was made by Sir Charles Parsons in 1919 at a meeting of the British Association for the Advancement of Science. Originally he proposed a very deep shaft interconnected with mined galleries. Today's schemes are rather different and involve drilling two boreholes, which are interlinked by an artificially stimulated network of natural fractures.

Creating a man-made reservoir large enough to turn cold water into water hot enough to drive a turbine is no easy task when it is a few kilometres deep underground. While the boreholes are technologically feasible, thanks to the oil and gas industries, expanding natural fissures in rocks over a controlled area so that they are large enough to carry water, but not likely to absorb it, is not. For the last ten years, therefore, scientists all over the world have been trying to perfect a means of achieving this, because the success of any hot dry rock scheme rests on the creation of a good heat-transfer system.

OPPOSITE ABOVE LEFT AND RIGHT *Two schematic representations of a hot rock reservoir formed by a stimulation of the natural granite joints in Cornwall* OPPOSITE BELOW LEFT AND RIGHT *The work at the Camborne School of Mines, Cornwall*

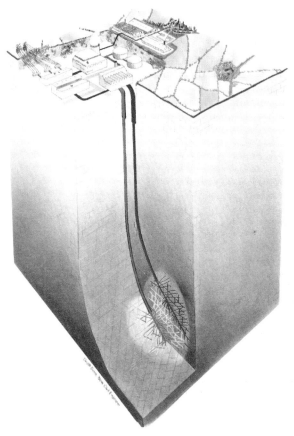

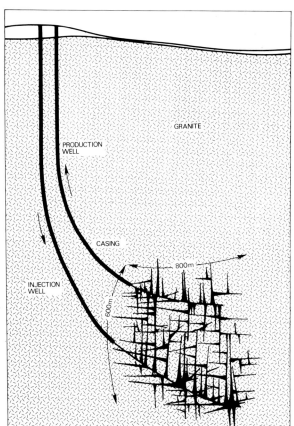

In theory hot dry rock systems are just that. They are dry, unlike the natural geothermal aquifers, and there is no fluid that can be pumped out. Instead water has to be injected down into the rock. By pumping cold water down one of the two boreholes slightly deeper than the other, the water is warmed up by the hot rocks. It passes through all the fissures until it reaches the second borehole where it arrives as superheated water, before being pumped up to the surface. Kept under sufficient pressure it will be prevented from boiling or flashing until it reaches the turbine to generate power in the standard way.

But the role of water in these systems goes further than that. It also plays a crucial part in interlinking the two boreholes. In Cornwall, where the trial wells have been drilled to a depth of 2000 m (6560 ft), the interconnection was achieved by pumping water at flow rates as high as 100 kilograms per second (220 lb per sec) – that is 6000 litres per minute (1320 gallons per min) and at a pressure of 140 kg cm$^{-2}$ (2000 psi). The rates have to be that high in order to ensure the maximum widening of the joints in the rock through which the water will flow. In Cornwall this has meant working with some of the greatest flow rates and durations that have ever been attempted. (At Camborne an explosive charge was fired at the bottom of the well to create a network of fractures at the borehole a good time before the water was injected.)

The reservoir created at Camborne was very large. By monitoring the microseismic activity during the water injection, the dimensions of the stimulated region became clear. It was at least 1500 m (4920 ft) high, 800 m (2625 ft) long and 300 m (985 ft) thick. The team had successfully produced a very large heat-transfer system indeed.

The work at Camborne, under the direction of Dr Tony Batchelor, is continuing and still has a considerable number of problems to face and stages to be reached before any geothermal-powered electricity results. One problem is the rate at which the water was absorbed by the reservoir: of the 300,000 m$^3$ (392, 400 cu yd) of water that was injected, only one-third was recovered in the test. Altering the depth at which they are operating will be another problem. For research and development purposes these tests are being carried out on boreholes 2000 m (6560 ft) deep where the temperature reaches 70°C. To generate electricity boreholes will have to be drilled to depths of between 6 and 8 km (3¾ and 5 miles), where temperatures are expected to reach 200°C.

It is now clear that hot dry rock technology has already taken huge steps forward; it has left the realms of theory and can now make a firm claim to be a future energy source. And since the Cornish granite area alone has been estimated to hold a greater energy potential than the current British coal reserves, the prize offered for mastering the technology is well worth seeking.

### Exploiting old oil wells

Why drill new boreholes when there are many unused or disused wells already in existence? That is the thought behind one British innovation which makes use of the many disused or unsuccessful oil and gas wells drilled under the sea or on land. The idea is to turn these wells into geothermal power stations, on the hot rocks principle. A heat pipe, 200 mm (7⁸⁄₁₀ in) in diameter, is pushed down a hole drilled deep into hot rocks. An organic fluid is then poured down the hole and boils at the bottom. The hot vapour produced rises back up through a separate tube inside the heat pipe. When it gets to the top the vapour gives up its heat, along the way driving a specially designed turbine. The condensed vapour goes back down the pipe to repeat the cycle. It is an ingenious idea, although as yet no practical schemes have yet been constructed. However, this idea of making hot rocks produce power is not the only way to use these geothermal resources. Hot rocks can also be used to get more mileage out of other available energy sources.

**Storing heat underground**

One of the main obstacles to the use of alternative energy sources like solar collectors for heating is the great time difference between when the heat is available and when it is needed. Any collector array large enough to supply midwinter heating needs would be grossly oversized for requirements during the rest of the year, and hopelessly uneconomic. This difficulty does not just apply to new technologies like solar collectors. Existing fossil fuel plants suffer the same problem: we always need a greater capacity than we use, so as to cope with the midwinter surge in demand.

Supposing we could use geothermal resources to act as an effective means of storing the energy collected during the spring and summer, when it does not have to be used, until it is needed in the winter? For now deep hole drilling, rock excavations and geophysical investigation techniques are playing a major role in enabling heat to be stored in rock for long periods, and to be extracted for district heating purposes when and where it is wanted. (The same methods can be applied in reverse: to store cold water for air-conditioning systems in hot climates.)

Large experimental systems are being built or tested in various parts of the world, but it is Sweden that a number of interesting projects have emerged. One is at the town of Luleå, which now hosts the world's first full-scale borehole heat store. The store consists of a large area of rock of 100,000 m³ (130,800 cu yd), perforated by 120 vertical boreholes, each 150 mm (6 in) in diameter and 65 m (213 ft) deep. The heat comes from surplus gas produced at a nearby steel plant. It is piped in and stored in the heat-retaining rock, which is heated by circulating hot water in the boreholes. The boreholes act as heat exchangers while heat is being stored and recovered.

The Luleå project was launched in 1982 and presently supplies heat for one of the buildings at the University of Technology there. The aim is to evaluate building techniques and the performance of the store for future designs and preliminary results are encouraging. During the first three years an average of 3600 megawatt hours of energy was stored, and an average of 2000 megawatt hours was recovered.

Another Swedish hole-drilling project is under way near the university city of Lund, but this time at much greater depths. In Gothenburg two large-diameter test holes have been drilled to depths of 750–800 m (2460–2650 ft) to determine whether the required water flows in the porous sandstone formations beneath the city could be achieved, and what kind of minerals and salt deposits would occur. For the project is in fact very similar to the hot dry rock experiment.

At the moment two full-scale production holes and two re-injection holes should be in operation by early 1985, and a maximum of 10 to 12 wells are due for completion some time after that date. Water will emerge at

250°C and be fed to industrial-scale heat pumps, through which most of the heat in water will be recovered. The pump installation will then distribute water heated to 80°C to homes and public buildings via the existing central-heating network.

### Energy in a cavern

Storing oil in large, excavated rock caverns may sound a little odd, but it is a technique that has been employed in Sweden for years. Now, it has been adapted to storing hot water instead. One of the most interesting examples is the Lyckebo project on the outskirts of the university city of Uppsala, where a newly blasted rock cavern with a volume of 100,000 m$^3$ (130,800 cu yd) has been filled with water. This will be heated during the summer to about 90°C to supply energy to the district's central-heating system during winter.

As the water is used over the winter the temperatures will drop, reaching about 40°C by the spring. The physiological qualities of the scheme mean different levels with different water temperatures will also be created. Instead of this being a disadvantage, the plant's engineers have turned it into a useful resource and exploited the differences. By allowing the water to travel in or out at all levels, water can be established or extracted at any temperature from any level to suit the circumstances. For example, the 90°C water in the upper part of the cave is retained for the winter, when the heating requirements are greatest. To ensure that they get the best performance and economy out of the storage plant, the engineers have installed a computerized control and regulation system. So if the 5000 megawatt hour cavern does fail to deliver, it will only be the computer that will be looking for excuses.

The success of the technique will be seen in the next few years; but such is the level of optimism that a full-scale project is under way in an abandoned mine, and in a lake water is stored in an abandoned power station tunnel. A number of projects are emerging which are directly coupled to the use of solar collectors as well, and the Lyckebo scheme just mentioned is eventually expected to be entirely heated by this method.

The use of the earth's heat both to provide energy and to store it has emerged within the last decade as an interesting, long-term prospect for the future. While there are only a finite number of natural geothermal locations, the potential offered if we overcome the difficulties involved in creating man-made reservoirs in hot dry rocks is very substantial indeed. If the development continues at the same rate then some at least of tomorrow's world may well be fuelled by geothermal heat and power.

---

# FURTHER READING

**British Wind Energy Association.** *Wind Energy for the Eighties*, Stevenage 1982. (*Windirections* (journal) and Conference Proceedings), Cambridge
**Commission of the European Communities.** *Acid Deposition: A Challenge for Europe*, Brussels 1983
**Cottrell, A.** *How Safe Is Nuclear Energy?*, London and New York 1981
**Count, B.M.** (editor). *Power from Sea Waves*, London and San Diego, Calif., 1980
**Elkington, J.** *Sun Traps: The Renewable Energy Forecast*, London and New York 1984
**Elsworth, S.** *Acid Rain*, London 1984
**Flood, M.** *Solar Prospects: The Potential for Renewable Energy*, London 1983
**Hoyle, F.** *Energy or Extinction?: The Case for Nuclear Energy*, London and New York 1977

**Hunt, S.E.** *Fission, Fusion and the Energy Crisis*, Oxford and Elmsford, N.Y., 1980
**National Radiological Protection Board.** *Living with Radiation*, London 1981
**Patterson, W.** *Nuclear Power*, 2nd edn, London and New York 1983
**Ramage, J.** *Energy: A Guidebook*, London and New York 1983
**Taylor, R.H.** *Alternative Energy Sources: For the Centralized Generation of Electricity*, Bristol and Philadelphia, Pa., 1983
**Warren Spring Laboratory.** *Acid Deposition in the United Kingdom*, London 1983

Numerous leaflets and pamphlets are also available free from the nuclear power industry, and from the Department of Energy in Britain and equivalent ministries and departments in other countries.

# INDEX ·